5th edition

geog.1

geography for key stage 3

< rosemarie gallagher >
< richard parish > < janet williamson >

OXFORD
UNIVERSITY PRESS

Great Clarendon Street, Oxford, OX2 6DP, United Kingdom

Oxford University Press is a department of the University of Oxford. It furthers the University's objective of excellence in research, scholarship, and education by publishing worldwide. Oxford is a registered trade mark of Oxford University Press in the UK and in certain other countries

© RoseMarie Gallagher, Richard Parish, Janet Williamson 2019

The moral rights of the authors have been asserted

First published in 2000
Second edition 2005
Third edition 2008
Fourth edition 2014
Fifth edition 2019

All rights reserved. No part of this publication may be reproduced, stored in a retrieval system, or transmitted, in any form or by any means, without the prior permission in writing of Oxford University Press, or as expressly permitted by law, by licence or under terms agreed with the appropriate reprographics rights organization. Enquiries concerning reproduction outside the scope of the above should be sent to the Rights Department, Oxford University Press, at the address above.

You must not circulate this work in any other form and you must impose this same condition on any acquirer

British Library Cataloguing in Publication Data
Data available

ISBN: 978-0-19-844604 0

11

The manufacturing process conforms to the environmental regulations of the country of origin.

Printed and bound in Great Britain by Bell & Bain Ltd, Glasgow

Acknowledgements

The publishers would like to thank the following for permissions to use copyright material:

Cover: Shutterstock/OUP; **p5**: NASA; **p6(tl)**: KEVIN A HORGAN/SCIENCE PHOTO LIBRARY; **p6(tr)**: Shutterstock; **p6(bl)**: travellinglight/Alamy; **p6(br)**: Robert Bannister/Alamy; **p7(t)**: Getty Images News/Getty Images; **p7(b)**: Shutterstock; **p8(tl)**: Nik Taylor/Alamy; **p8(tr)**: Shutterstock; **p8(bl)**: © Bryan & Cherry Alexander Photography/Arctic Photo; **p8(br)**: nik wheeler/Getty Images; **p9(t)**: NASA; **p9(m)**: Shutterstock/NASA; **p11(l)**: Lanmas/Alamy; **p11(m)**: Ian G Dagnall/Alamy; **p11(r)**: Royal Geographical Society/Getty Images; **p11(bl)**: Kumar Sriskandan/Alamy; **p11(br)**: Shutterstock; **p13**: Midland Aerial Pictures/Alamy; **p14(t)**: BRIAN HARRIS/Alamy; **p14(m)**: Ironbridge Gorge Museums Trust; **p14(b)**: Robin Weaver/Alamy; **p17(tl)**: Malcolm Couzens; **p17(tr, b)**: David Bagnall/Alamy; **p20**: NASA; **p21**: Granger Historical Picture Archive/Alamy; **p22(t)**: Art Heritage/Alamy; **p22(ml)**: Science History Images/Alamy; **p22(mr)**: North Wind Picture Archives/Alamy; **p22(b)**: Coston Stock/Alamy; **p23(t)**: Mitchell Library, State Library of New South Wales; **p23(b)**: Robert Ashton/Massive Pixels/Alamy; **p24(l, r)**: Shutterstock; **p25**: Shutterstock; **p26**: Oxford University Press; **p28**, **30**: Getmapping.com; **p33(t)**: Greg Balfour Evans/Alamy; **p33(b)**: Tony Watson/Alamy; **p35**: © Crown copyright 2019 OS100000249; **p36(t)**: Cotswolds Photo Library/Alamy; **p36(t)**: keith morris/Alamy; **p36(b)**: © Crown copyright 2019 OS100000249; **p38**: DAVID VAUGHAN/SCIENCE PHOTO LIBRARY; **p40(l)**: Vanda Ralevska/Alamy; **p40(r)**: RosaIreneBetancourt 6/Alamy; **p41(tl)**: Pawel Libera/Getty Images; **p41(tr)**: Andrew Fox/Alamy; **p41(bl)**: Loop Images Ltd/Alamy; **p41(br)**: James Morris/Alamy; **p43(l)**: www.mjt.photography/Alamy; **p43(r)**: BANANA PANCAKE/Alamy; **p49**: Sally and Richard Greenhill/Alamy; **p51(l)**: NiKreative/Alamy; **p51(r)**: A.P.S. (UK)/Alamy; **p53(l)**: Shutterstock; **p53(r)**: Nick Ansell/PA Archive/PA Images; **p54(t)**: ESA/NASA (want credit in caption, will discuss with them); **p54(b)**: Jason Bryan/Alamy; **p55(l)**: imageBROKER/Alamy; **p55(r)**: Dylan Garcia/Alamy; **p56**: Bloomberg/Getty Images; **p57(t)**: MANAN VATSYAYANA/Getty Images; **p57(b)**: Russell Kord/Alamy; **p59**: J Oerlemans/Utrecht University; **p60(l)**: Tony Waltham; **p60(r)**: Shutterstock; **p61**: dpa picture alliance archive/Alamy; **p63(t)**: Juerg Alean, Eglisau, Switzerland; **p63(l)**: NASA; **p63(br)**: Don B. Stevenson/Alamy; **p64(t)**: MICHAEL P. GADOMSKI/SCIENCE PHOTO LIBRARY; **p64(b)**: J Oerlemans/Utrecht University; **p65**: J Oerlemans/Utrecht University; **p66(l)**: With kind permission from Alan Green; **p66(r)**: Michael Graham/Bleaberry Tarn/CC BY-SA 2.0; **p67(t)**: Robertharding/Alamy; **p67(b)**: Shutterstock; **p68(l)**: Paul Heinrich/Alamy; **p68(r)**: Andrew Leaney; **p69(l)**: Linda Lyon/Getty Images; **p69(r)**: Shutterstock; **p70(l)**: With kind permission from Ann Bowker; **p70(r)**: Shutterstock; **p71(t)**: Ashley Cooper/Alamy; **p71(b)**: Tony Waltham; **p72(t)**: Photimageon/Alamy; **p72(b)**: © Acceleratorhams/Dreamstime; **p73**: © Crown copyright 2019 OS100000249; **p74(tl)**: Shutterstock; **p74(tr)**: Tony Waltham; **p74(b)**: Agencja Fotograficzna Caro/Alamy; **p75(tl)**: Galen Rowell/Mountain Light/Alamy; **p75(tr)**: ARCTIC IMAGES/Alamy; **p75(b)**: Robert Wallis/Panos; **p76(tl)**: Peter Darch/Alamy; **p76(tr)**: © Crown copyright 2019 OS100000249; **p76(b)**: Pearl Bucknall/Alamy; **p77**: Shutterstock; **p78(tl)**: Nature Photographers Ltd/Alamy; **p78(tr)**: Colin Underhill/Alamy; **p78(ml)**: robertharding/Alamy; **p78(mr)**: Brian Shaw/Alamy; **p78(bl)**: Omar Farooque; **p78(br)**: Shutterstock; **p79**: Omar Farooque; **p80**: Shutterstock; **p84**: Oonat/Getty Images; **p85**: Kathy DeWitt/Alamy; **p86(t)**: Partrick Mackie/www.geograph.org.uk/CC BY-SA 2.0; **p86(b)**: David Norton/Alamy; **p87**: Heather Angel/Natural Visions; **p88(t)**: Stephen Dorey ABIPP/Alamy; **p88(ml)**: Photofusion Picture Library/Alamy; **p88(m)**: Roger Bamber/Alamy; **p88(mr)**: Lars Plougmann/Wikimedia; **p88(bl)**: Gary K Smith/Alamy; **p88(bm)**: Scott Campbell/Alamy; **p88(br)**: John Eccles/Alamy; **p89(t)**: KEVIN ELSBY/Alamy; **p89(m)**: Shutterstock; **p89(b)**: geogphotos/Alamy; **p90(t)**: Nigel Sawyer/Alamy; **p90(m)**: Andi Edwards/Alamy; **p90(b)**: © ZSL Jonathan Kemeys; **p91(tl)**: Frank Watson; **p91(ml)**: James Bell/Alamy; **p91(bm)**: Thurrock Thameside Nature Park; **p91(tr)**: Paul White Aerial views/Alamy; **p91(br)**: Tony Watson/Alamy; **p94(tl)**: Toby Melville/Reuters; **p94(tr)**: Fraser Pithie/Alamy; **p94(mr)**: Peter Macdiarmid/Getty Images; **p94(bl)**: Michael Lloyd/INS News Agency; **p94(br)**: Peter Macdiarmid/Getty Images; **p95**: © Crown copyright 2019 OS100000249; **p96(tl)**: Iain McGillivray/Alamy; **p96(tr)**: Vagner Vidal /INS News Agency; **p96(bl)**: RTimages/Alamy; **p96(br)**: Jeremy Moeran/Alamy; **p97(tr)**: Environment Agency; **p97(tr)**: Alessia Pierdomenico/REUTERS; **p97(b)**: Hoberman Collection/Alamy; **p98(t)**: Education Images/Getty Images; **p98(b)**: Steve Benson MTC www.projectmirfield.co.uk; **p99**: NASA; **p101(t, b)**, **104(t, m)**: Shutterstock; **p104(bl)**: Jake Lyell/Alamy; **p104(br)**: Frank Kahts/Alamy; **p105**: BSIP/UIG/Getty Images; **p107(t)**: Rolf Richardson/Alamy; **p107(b)**: Photo 12/Alamy; **p109(l)**: Shutterstock; **p109(r)**: Sven Torfinn/Panos; **p111(tl)**: Shutterstock; **p111(tr)**: frans lemmens/Alamy; **p111(b)**: Danita Delimont/Getty Images; **p112(t)**: MELBA PHOTO AGENCY/Alamy; **p112(b)**: Ariadne Van Zandbergen/Alamy; **p113(l)**: Shutterstock; **p113(r)**: Eric Nathan/Alamy; **p115**: Ariadne Van Zandbergen/Alamy; **p116(t)**: Shutterstock; **p116(m)**: Peter Treanor/Alamy; **p116(b)**: Avalon/UIG/Getty Images; **p117(tl)**: Shutterstock; **p117(tr)**: Nina Hassa/Alamy; **p117(b)**: Joerg Boethling/Alamy; **p118**: Neil Thomas/Getty Images; **p119(t)**: John Warburton-Lee Photography/Alamy; **p119(m)**: Martin Harvey/Alamy; **p119(bl)**: Shutterstock; **p119(br)**: Images of Africa Photobank/Alamy; **p121(tl)**: Images of Africa Photobank/Alamy; **p121(tr)**: Marion Kaplan/Alamy; **p121(b)**: Shutterstock; **p124**: Marja_Schwartz/iStockphoto; **p126**: Sueddeutsche Zeitung Photo/Alamy; **p127(1-5)**: Shutterstock; **p127(6)**: TONY KARUMBA/Getty Images; **p127(7)**: Dbimages/Alamy; **p127(8)**: Dbimages/Alamy; **p127(9)**: Jan Hetfleisch/Shutterstock; **p128(t)**: Wayne HUTCHINSON/Alamy; **p128(m)**: Lake Turkana Wind Power; **p128(b)**: Shutterstock; **p129(tl)**: Dorling Kindersley ltd/Alamy; **p129(tr)**: Mobius Motors; **p129(b)**: Benedicte Desrus/Alamy; **p130(l)**: AFP/Getty Images; **p130(m)**: Shutterstock; **p130(r)**: Kumar Sriskandan/Alamy; **p131(l)**: Mark Boulton/Alamy; **p131(m)**: Shutterstock; **p131(r)**: Mark Boulton/Alamy; **p132(tl)**: YASUYOSHI CHIBA/Getty images; **p132(tr)**: Avalon/Photoshot License/Alamy ; **p132(bl)**: David Tyrer/Alamy; **p132(br)**: Shutterstock; **p133**: imageBROKER/Alamy; **p134(t)**: Sven Torfinn/Panos; **p134(b)**: Wayne HUTCHINSON/Alamy; **p135**: John Warburton-Lee Photography/Alamy.

Artwork by: Mike Phillips (p10, 12-13, 102-3, 122-3), Ian West, QBS Media Services Inc., Mike Parsons, and Simon Tegg.

Ordnance Survey (OS) is the national mapping agency for Great Britain, and a world-leading geospatial data and technology organisation. As a reliable partner to government, business and citizens across Britain and the world, OS helps its customers in virtually all sectors improve quality of life.

The changes in this edition of geog.1 are the result of comments from many people. We would like to thank the teachers who came together in focus groups to discuss the course, as well as our reviewers who provided thoughtful and constructive criticism. Thanks to Kate Stockings, and to Richard Aldred and Gillian Crumpton of the Ironbridge Gorge Museum Trust. A special thanks to Garaeth Davies.

Note that the content of any direct speech attributed to characters in this book is based on information from reliable sources. Walter and Violet in Chapter 2 are based on students we know.

Although we have made every effort to trace and contact all copyright holders before publication this has not been possible in all cases. If notified, the publisher will rectify any errors or omissions at the earliest opportunity.

Links to third party websites are provided by Oxford in good faith and for information only. Oxford disclaims any responsibility for the materials contained in any third party website referenced in this work.

Contents

1	**Geography … and you**	**5**
1.1	Welcome to geography!	6
1.2	What's in your geography kit?	8
1.3	How to get good at geography	10
1.4	Change in the Ironbridge Gorge	12
1.5	How to answer questions – part 1	14
1.6	How to answer questions – part 2	16
1.7	How to answer questions – part 3	18
	Geography … and you Check	20

2	**Maps and mapping**	**21**
2.1	Mapping through the ages	22
2.2	Plans and scale	24
2.3	The maps in your head	26
2.4	From an aerial photo to a map	28
2.5	Using grid references	30
2.6	How far?	32
2.7	Ordnance Survey maps	34
2.8	How high?	36
2.9	Where on Earth?	38
	Maps and mapping Check	40

3	**About the UK**	**41**
3.1	Your island home	42
3.2	It's a jigsaw!	44
3.3	What's our weather like?	46
3.4	Who are we?	48
3.5	Where do we live?	50
3.6	How are we doing?	52
3.7	London, our capital city	54
3.8	Our links to the wider world	56
	About the UK Check	58

4	**Glaciers**	**59**
4.1	Your place … 20 000 years ago!	60
4.2	Glaciers: what and where?	62
4.3	How do glaciers shape the land?	64
4.4	Landforms shaped by erosion – part 1	66
4.5	Landforms shaped by erosion – part 2	68
4.6	Landforms created by deposition	70
4.7	More about the Lake District	72
4.8	Do glaciers matter?	74
	Glaciers Check	76

5	**Rivers**	**77**
5.1	Meet the River Thames	78
5.2	It's the water cycle at work	80
5.3	A closer look at a river	82
5.4	How do rivers shape the land?	84
5.5	Six landforms created by rivers	86
5.6	How do we use rivers?	88
5.7	What's the Thames Estuary like?	90
5.8	Floods!	92
5.9	Flooding on the River Thames	94
5.10	Can we protect ourselves from floods?	96
	Rivers Check	98

6	**Africa**	**99**
6.1	What and where is Africa?	100
6.2	A little history	102
6.3	What's Africa like today?	104
6.4	The countries in Africa	106
6.5	Population distribution in Africa	108
6.6	What are Africa's main physical features?	110
6.7	Africa's biomes	112
	Africa Check	114

7	**Kenya**	**115**
7.1	Hello Kenya!	116
7.2	What are Kenya's main physical features?	118
7.3	What's Kenya's climate like?	120
7.4	A short history of Kenya	122
7.5	Kenya's population	124
7.6	What's Nairobi like?	126
7.7	What does everyone do?	128
7.8	How Kenya earns money from flowers	130
7.9	On safari!	132
7.10	So how is Kenya doing?	134
	Kenya Check	136

Command words: a summary	137
Key for OS maps	138
Map of the British Isles	139
Map of the world	140
Glossary	142
Index	144

Where are you?

1 Geography ... and you

1.1 Welcome to geography!

 This unit introduces geography, and the kinds of things you will learn about. Lucky you!

Glorious geography

Geography is about our planet and us. It is linked to almost everything in our lives. It will help you to understand your world – and to shape it.

Dividing up geography

Geography is a big subject. So we divide it into different strands. You'll study the three strands shown here.

1 Physical geography – what our planet is like

You'll learn about oceans, rivers, glaciers, the coast, weather, climate, rocks. How Earth's surface is made of big slabs. And about the dangers we face from Earth – earthquakes, erupting volcanoes, and more.

2 Human geography – how and where we live

You'll learn about how our numbers are growing. Where we live, and how we earn a living. You'll study some countries in quite a lot of detail … and see how and why some are poorer than others.

3 Environmental geography – our impact on our surroundings

We share Earth with other living things … but we're squeezing them out. We pollute the air and ocean and soil. We are making Earth warmer. What can we do? You'll think about all this!

Three big themes in geography

Here are three big themes that run through geography.

1 Change

All over Earth there is continual **change**. Look at these examples.

- The big slabs that form Earth's hard shell are sliding around, causing earthquakes and volcanic eruptions. Rivers are wearing land away. Rocks are breaking down to soil.
- The number of humans is changing too. In the year 1800 there were about 1 billion of us. Today, we are over 7 billion.

2 Impact

All change has an **impact** on someone, or something. For better or worse.

- Earth provides everything we need – water, soil to grow food, fuels, metals, and more.
- But Earth can harm us too. For example through earthquakes and floods.
- Our actions have an impact on Earth and on other species. We litter rivers and oceans with plastic waste. Experts say we are warming Earth up, by burning fossil fuels – coal, oil, gas. (We are trying to do less harm!)

3 Inequality

There is huge **inequality** among us humans.

- Almost half of us live in **poverty**, on less than £4.50 a day. For food, water, fuel, a place to stay, clothing, medicine, shoes, school fees, everything. Hundreds of millions are in **extreme poverty**, with almost nothing.
- And think about this. Around 60 of the world's richest people own the same wealth as 3.7 billion of the world's poorest people. (That's 3 700 000 000 poor people.)

Look out for these themes as you work through your course.

▲ When change means danger! Lava from an erupting volcano runs along a road in Hawaii.

▲ A fish swimming among plastic waste. Fish eat plastic – they mistake it for food.

Your turn

1. Copy and complete, to make a sentence:
   Physical geography is about …
   Human geography is about …
   Environmental geography is about …

2. In which strand of geography might you learn about this?
   - a how clouds form
   - b the world's biggest cities
   - c protecting pandas
   - d where trainers are made
   - e mountains
   - f plastic litter in the ocean

3. a List three big themes that you'll meet in geography.
   b We have an *impact* on Earth. Explain what that means.
   c What does *inequality* mean? (Glossary?)

4. One big theme in geography is *change*. Give one example of a *natural* change on Earth (not caused by humans).

5. Experts think that we humans *(homo sapiens)* first appeared around 315 000 years ago, in Africa. Write a list of ways we have changed Earth's surface since then.. (For example we have cut down forests … built … mined …)

6. Billions of people live on less than £4.50 a day, for everything. Name something you buy, that costs about £4.50.

7. Look at these three questions. Choose one – and answer it!

a Who owns Earth anyway?
b Why bother looking after Earth?
c Why does it matter if people are poor?

1.2 What's in your geography kit?

 This is about the resources you will use, in studying geography.

Exploring your planet

When you do geography, you're an explorer! These are the resources you will use for exploring – mostly without leaving your desk.

1 Maps

Maps are top of the list. They are drawings of places as if from above.

They can show huge areas – like the whole world. Or small areas, like the streets where you live. They can show **themes**, like the amount of rain that falls on places.

You'll find maps everywhere. In parks, on your computer, on your mobile. And in an **atlas**.

▶ Maps are everywhere.

2 A compass

A **compass** tells you direction.

But you don't need to carry one. You can work out north, south, east, and west in your head. (The sun rises in the east and sinks in the west.)

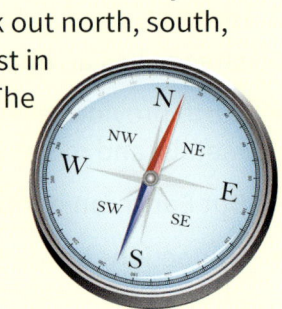

▲ A compass needle swings to show north.

3 Photographs

▲ A young reindeer herder in Siberia in Russia. Her family lives in a tent and moves with the reindeer to find pasture.

Photos are a great way to explore a place.

The photo above was taken at ground level. The one on the right is an **aerial photo**.

Aerial photos are taken from above – using cameras attached to planes, or balloons, or drones.

▲ The Marsh Arabs live on islands of reeds in the marshes of southern Iraq, and go everywhere by boat. Their way of life is under threat.

Geography... and you

4 Satellite images

Satellite images are taken from high above Earth by **sensors** in satellites. They can cover large areas.

They are used for all kinds of things. For example to track hurricanes and wildfires. Or even to spy. You could spot signs that a country was building nuclear weapons, for example. (Some sensors can even pick out dustbins.)

▶ *Around 700 satellites orbit Earth capturing images. A full orbit from pole to pole can take as little as 1.5 hours. (Hundreds of other satellites do other jobs.)*

5 Google Earth

Google Earth is a program you download from the internet. It combines maps, satellite images, photos, and 3D images.

With Google Earth you can travel through cities, towns, and villages. Zoom in for a closer look. Follow rivers. Cross deserts. Visit mountains and valleys on the ocean floor.

Try it! You could start by looking for your home.

6 GIS

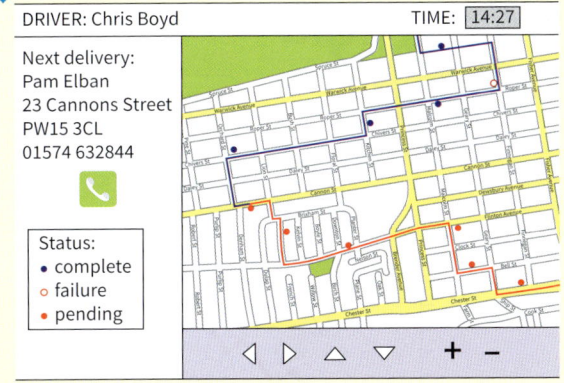

A **GIS**, or **geographic information system**, is used to display and analyse data. It is based on maps.

Suppose you want to do a survey of cafés near school. You can collect data about each café, including its latitude and longitude (or postcode). Then feed the data into a GIS program. The cafés will show up on a map on your screen.

GIS is used to help people make decisions. For example to help drivers plan the best route when delivering parcels. Or to help police identify areas that are hotspots for crime.

There is more on GIS in *geog.2*.

▲ *Where next? GIS helping a delivery driver.*

Your turn

1. What is: **a** a map? **b** an aerial photo?

2. You have to find your way around a new town. Which do you think will be the most helpful?
   a a map **b** a compass **c** photos
   Explain your answer.

3. You are trying to decide which new town to visit, on holiday. Which will be more helpful in making up your mind?
   a a map **b** photos
   Explain your answer.

4. Your job is to check continually for illegal **logging** (chopping down trees) in Brazil's rainforest. Which might be more useful, and why?
   a aerial photos **b** satellite images

5. Give three examples of what you can use Google Earth for.

6. What is a *GIS*, and what is it used for?

7. You want to explore a country from your desk. Choose three tools from this unit to help you. List them in order, the most useful one first. Write a paragraph to explain your choice.

1.3 How to get good at geography

 You can get really good at geography. Let's see how.

1 Be an explorer!

Doing geography means you are an **explorer** of your planet.

Explorers are adventurous, and full of curiosity. They have a plan. They know where they're heading and are determined to get there. They are prepared.

So think like an explorer. And be prepared. Have your **geography kit** ready and know how to use it. (You met it in Unit 1.2.)

You can explore a lot from your desk. But later, when you do **fieldwork**, you'll explore outdoors too.

2 Don't get lost!

Exploring is fun – but not if you get lost!

Knowing where places are is key in geography. If you are exploring Mount Kilimanjaro, you should know it's in Tanzania, in Africa. So pay attention to **location**. Find out where places are. An atlas will help.

▲ The geography explorers set out.

3 Ask lots of questions

As an explorer, you must get nosy.
Ask lots of questions about the places you explore.
Questions that begin with:
Who …
What …
Where …
How …
Why …
When …
And then look for clues, to find the answers.

4 Be ready for some maths

Lots of things are **measured** and **counted**, in geography. Like rainfall, temperature, latitude, population, wealth.

So you will use numbers from time to time. You'll look up data in tables. You'll interpret bar charts, and pie charts, and line graphs. You'll look for patterns and trends.

You learn these skills in maths class. Using them in geography is fun.

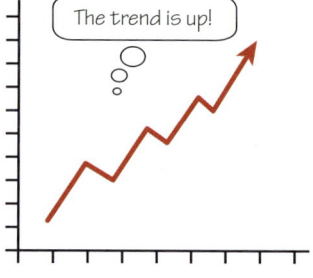

5 Be ace at writing answers

All through your geography course, you'll write answers to questions.

The questions usually contain **command words** that tell you how to answer them. Words like *describe, calculate, explain*.

Units 1.5 – 1.7 are all about command words. They will help you. And there's a list of command words on page 137.

▲ Ibn Battuta (Moroccan, 1304 – 1368).
He explored widely for thirty years – mainly in Asia – and wrote about his adventures.

▲ Captain Cook (British, 1728 – 1779).
He sailed thousands of miles in the Pacific Ocean, exploring and mapping places.

▲ Freya Stark (British, 1893 – 1993).
She explored and mapped areas in the Middle East, and wrote over 20 travel books.

Your turn

1 Three explorers are shown above.
  Why were they willing to risk their lives to explore places? See if you can think up some reasons.

2 Geographers always want to know where places and features are. Where are these? Give the country and continent.
  a Paris
  b the River Ganges
  c Hollywood
  d the Great Barrier Reef
  e Mount Snowdon
  f the Atacama desert

3 Photo **A** below shows people on holiday. Get nosy!
  a Why might people choose this place for a holiday? List as many reasons as you can.
  b Why is there snow here?
  c How might this place change, later in the year?

4 Study photo **B**. Then answer these questions.
  a What is going on in the photo?
  b Why is the place like this?
  c Who do you think is responsible?

5 a Now make up three new questions about photo **B**, and what's going on there. No silly ones!
     (Hint: Who? What? Where? How? Why? When?)
  b Ask your partner to try to answer them.

6 *Change* is a big theme in geography.
  Think of a change going on in your area. (For example, a new road being built, or flats being knocked down.)
  Make up questions about the change, like those in section 3 on page 10. Then see how many you can answer.

A

B

1.4 Change in the Ironbridge Gorge

 As you saw in Unit 1.1, change is a big theme in geography. Here we look at an example of change in Shropshire.

The story of the Ironbridge Gorge

The Ironbridge Gorge is an area in Shropshire in England. It lies along a steep valley or **gorge** where the River Severn flows. Look at map **A**. Photo **B** shows part of the area. Today it attracts many tourists. Why? Let's see.

People first settled on the slopes of the gorge in the 8th century. They farmed, and used wood for building and fuel, and the river for transport.

The area had other **natural resources** too: coal, limestone, iron ore, and different clays. These were exposed on the slopes, and easy to dig out.

So people began to use coal for fuel, and limestone for building, and clays to make pottery and clay pipes. Things could be moved by boat, to sell.

They also extracted iron from iron ore, by heating the ore up with limestone and charcoal made from wood. Trees were cut down to get the wood.

But in 1709 Abraham Darby found a way to extract iron using coke from coal, instead of charcoal. A big step towards the Industrial Revolution!

For the rest of the 18th century the local iron industry thrived. Making iron wheels. Rails. Cylinders. Pots and pans. And lots of smoke and grime.

In 1781 a cast iron bridge was opened across the river, the first of its kind in the world. A town grew beside it. Its name: Ironbridge! (See photo **B**.)

But there was a drawback. The steep slopes of the gorge meant it was not so easy to access the area, or for industries and towns to grow.

So during the 19th century the area began to decline – while other places, with the same resources and easier access, were doing well.

By the 1960s most industries had gone. But people had a bright idea: to preserve the abandoned sites, and celebrate their history.

Today the Ironbridge Gorge has ten museums, including an open-air one. They tell about its industrial past, and how people lived and worked.

Now there is no smoke and grime. The area's main industry is **tourism**. And many school groups visit, to learn more about Britain's past.

The Ironbridge Gorge is only one example of how people change a place. Change happens everywhere – including where you live!

▲ Shropshire. Find the Ironbridge Gorge.

▲ Ironbridge, a small town in the Ironbridge Gorge. Named after …?

Your turn

1. What is a *gorge*? (Glossary?)

2. Where is the Ironbridge Gorge? Give your answer as a paragraph. Include the names of:
   - the nation and county it's in
   - the river that runs through it
   - the town nearest to it on the map above

3. Was the Ironbridge Gorge always called this? Explain.

4. a What is a *natural resource*? (Glossary?)
   b List five natural resources found in the Ironbridge Gorge. (Is soil a natural resource? Is a river?)
   c Describe how people used each resource you listed in **b**, in the past. (A resource may have more than one use.)

5. a Name the man who found a way to extract iron from its ore commercially, using coke made from coal.
   b This development was a big step towards the *LAISTURIND VOONITURLE*. Unjumble the term in italics and then explain it.

6. The steep slopes of the Ironbridge Gorge made it easy to see and dig out coal and limestone and iron ore and clay. But the slopes were a disadvantage too.
   a In what way were they a disadvantage?
   b How did this disadvantage affect industry there?

7. Write a paragraph to describe how people have changed the Ironbridge Gorge, from the 8th century up till today.

13

1.5 How to answer questions – part 1

This is the first of three units to help you answer questions in geography. It is based on the Ironbridge Gorge.

What are command words?

There are lots of questions for you in *geog.123*. Many of them contain **command words**. These words tell you how to answer the question. So be smart. Look for the command words first. Then do what they say!

Your first set of command words
These are in alphabetical order, to help you.

Calculate
Means Do some maths, to get the answer!
Always give the unit. For example 5 *km* or 11 *people* or 15 *days*.
See question 4 in *Your turn*.

Copy and complete
Means Copy this, filling in all the blanks.
See question 2 in *Your turn*.

Define
Means Write down the meaning.
Keep your answer clear and simple.
See question 1 in *Your turn*.

Give
Means Come up with an answer, from what you have learned already.
Again, keep it clear and simple.
See question 4 in *Your turn*.

Identify
Means Pick out this thing, and give its name.
See question 3 in *Your turn*.

Name
Means Write the name of the thing you are asked about.
Easy! You do not need to write a full sentence.
See question 3 in *Your turn*.

State
Means Give the answer in clear terms.
It is often used in place of *Give* or *Identify* or even *Calculate* or *Count*.
Just be sure to answer clearly.
See question 4 in *Your turn*.

Suggest
Means Come up with a possible answer.
You may not find the answer in the text. So use your common sense!
See question 6 in *Your turn*.

Now apply these command words in *Your turn*.

▲ The famous Iron Bridge, opened in 1781. People paid a toll to use it. Before this, bridges were made of wood or stone.

▲ Blists Hill in the Ironbridge Gorge is an open-air museum of streets, shops, workshops and homes – as they were in Victorian times.

▲ This is Enginuity, another museum in the Ironbridge Gorge. It shows off innovation, linking the past, present and future.

Geography... and you

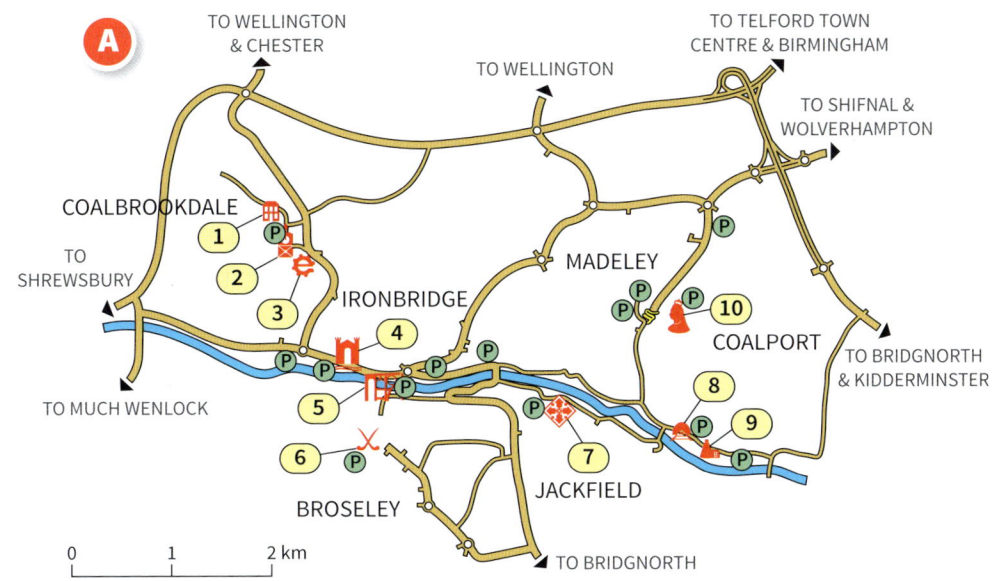

▲ The Ironbridge Gorge has several towns and villages. And ten museums celebrating its past history.

Your turn

1. All the questions in this unit contain command words. **Define** the term *command words*.

2. **Copy and complete** this paragraph about the Ironbridge Gorge, using words from the list in brackets.

   The Ironbridge Gorge is often called the _____ of the Industrial _____. Its iron _____ was very important, but so were others. For example its _____ and _____, and clay _____ for smoking tobacco, were _____ around the world. Today you can explore this _____ past in the _____ of the Ironbridge Gorge.

   (china   museums   Revolution   industry   birthplace   tiles   sold   industrial   pipes)

3. Map **A** shows that the Ironbridge Gorge has other small towns and villages besides Ironbridge.
   a **Name** any three of them.
   b The names of two villages hint that coal was found there. **Identify** these two villages.

4. Map **A** shows the Ironbridge Gorge museums.
   a **Give** one reason why tourists visit the Ironbridge Gorge.
   b **Name** three museums in the Ironbridge Gorge.
   c **State** the number of parking areas shown on map **A**.
   d Enginuity is 2.3 kilometres by road from the iron bridge, which is 4.3 km from the China museum. **Calculate** the distance from Enginuity to the China museum.

5. This bar chart shows visitor numbers to the Ironbridge Gorge museums in a recent year. The numbers are in thousands.

   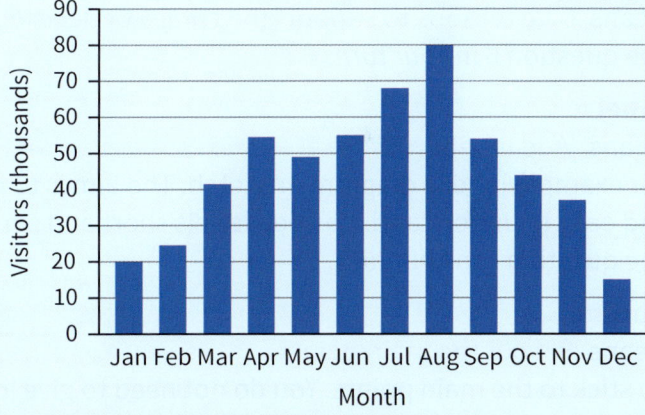

   a **Identify** the month with most visitors.
   b **Name** the months with more than 50 000 visitors.
   c **State** the number of visitors there were in:
      i January   ii August
   d **Calculate** how many more visitors there were in August than in January.

6. Look again at the bar chart above. **Suggest** two reasons why there were more visitors in August than in January.

7. Now make up four questions about your own area, using different command words from page 14. Swop them with a partner. Then answer each other's questions!

1.6 How to answer questions – part 2

 You already met command words in Unit 1.5. Here you will meet some more, to help you answer questions.

More command words for you

Below are more command words, to add to the ones you met already. Again they are in alphabetical order.

Compare
Means **Identify what is the same, and different, about two things.**
For example say which one is bigger. Mention *both* things in your answer.
See question 3 in *Your turn*.

Describe
Means **Write a description.**
For example describe what you see, or the steps in a process.
You do not need to give reasons for anything.
See question 2 in *Your turn*.

Draw
Means **Draw!**
For example draw a diagram, or a sketch map, or a bar chart, or a line graph. Use a ruler for straight lines. Try to be accurate – and quick!
See question 1 in *Your turn*.

Explain
Means **Give reasons why something is the way it is.**
See question 5 in *Your turn*.

Justify
Means **Give reasons to support the choice or decision you made.**
See question 6 in *Your turn*.

Label
Means **Add labels!**
For example label a diagram or sketch. The aim is to make it clear and easy to understand. So keep labels short and simple!
See question 1 in *Your turn*.

Outline
Means **Set out the main points.**
So stick to the main points. You do not need to give lots of detail.
See question 4 in *Your turn*.

Command words in everyday life

In fact you use or follow command words all the time, in everyday life. Look at strip **A**. These command words have the same meaning in geography. So you already know how to apply them!

A

▲ This road in Ironbridge is called **The Wharfage**. (A wharf is a place where boats are loaded and unloaded.) It runs beside the River Severn, near the iron bridge. See the map on page 15.

▲ The Wharfage at a different time! The River Severn has flooded it. These days when there is a flood warning, they put temporary barriers along The Wharfage to keep the water out.

Your turn

1. Ironbridge is on the River Severn. This is the UK's longest river. Its **source** (starting point) is in Wales. It ends its journey in England, in the Bristol Channel. Map **D** shows its **course** (the route it takes). Check page 139 too.
   a **Draw** your own map to show the course of the Severn.
   b On your map:
      i **label** the river, the Cambrian Mountains, Wales, England, the Bristol Channel, and the Atlantic Ocean
      ii add dots for the six places shown on map **D**, and **label** them.
      iii add notes to say how long the river is (354 km) and the height of its source above sea level (610 m).
   c Give your map a title.

2. **Describe** the course of the River Severn as clearly as you can. Write a paragraph with at least four sentences.
   - Don't forget to say where the river starts and ends its journey.
   - Does the course follow a smooth line? Does it change direction? Does its shape remind you of anything?
   - Add any other information that you think will improve your description. (Check your map from **1**.)
   - Write a title for your paragraph.

3. **Compare** photos **B** and **C**. What is the key change from **B** to **C**? Don't forget to name the street and river.

4. a **Suggest** one reason why the River Severn floods from time to time. (Think about weather!)
   b **Outline** one impact of flooding on the shops in **C**.

5. Photo **E** shows The Wharfage in a later year, when the River Severn was in flood. A steel barrier has been put up.
   a **Explain** why the steel barrier was put there.
   b **Compare** the heights of the woman and the barrier.
   c What evidence is there that the barrier is working?

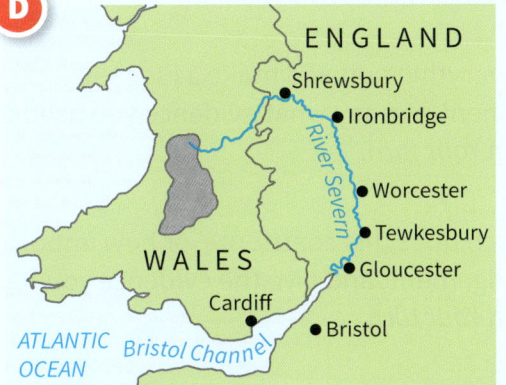

6. Think about this person's opinion.

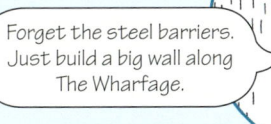

Forget the steel barriers. Just build a big wall along The Wharfage.

   a Do you agree with this opinion? Decide.
   b Then **justify** your decision. Give at least one reason!

7. Now make up *three* questions about your own area, using different command words from page 16. Swop them with a partner. Then answer each other's questions!

1.7 How to answer questions – part 3

 Here we look at our final set of command words. And again you'll see how they work in everyday life!

Command words that need longer answers

These need longer answers. Not just one word or one sentence!

Discuss
Means Look at this from different angles, and give key points about it.
See question 1 in *Your turn*.

Examine
Means Look at each part, and say how it contributes to the whole thing.
See question 4 in *Your turn*.

Evaluate
Means Judge how successful or worthwhile something is.
You should give its good and bad points, and your final opinion about it.
See question 3 in *Your turn*.

Assess
Means Weigh everything up and make a judgement.
Give your judgement and say what evidence you used to reach it.
See question 5 in *Your turn*.

To what extent .... ?
Means How much does this contribute, or how important or true is it?
Again make a judgement, and give the evidence you based it on.
See question 6 in *Your turn*.

How to write longer answers

The commands above have a lot in common.

- They expect you to know quite a lot about a topic, and to show this.
- They expect you to write about different aspects of the topic – for example what's good and what's bad about it. Give several facts.
- You have to weigh everything up and give a final opinion or judgement.

So here's an approach to writing the answers:

> Write an opening sentence that states what the topic or issue is.

> Then write sentences with the points you want to cover. Put them in a logical order. (For example the positive points first, then the negative ones.) Introduce the negative or opposing points with words like *But ...* or *However ...* or *On the other hand ...*

> Write a closing sentence giving your final opinion or judgement, in response to the question. Begin with words like *On balance ...* or *In conclusion ...* or *To summarise ...*

A

Discuss!
"We didn't play well." "Joe was awful." "Their goalie is ace." "We won't make the top four."

Examine!
"Light frame." "Wide wheels." "Padded seat." "The right height." "Perfect for me." "18 gears."

Evaluate!
"A good essay. Well written." "But poor spelling." "7 out of 10." "Kayo."

Assess!
"No, not for me." "Nice colours. Really cheap. But too thin. Too long. Too loose."

To what extent ...?
"So ... to what extent is your life ruled by social media?" "Well, I check every ten minutes when I'm awake."

Your turn

Geography... and you

Note: *for some of these questions you must use what you already know about the Ironbridge Gorge. You may need to look back at earlier units.*

1. Look at cartoon strip **A**. It shows five commands in practice in everyday life.
   a. Choose one of the commands, and give an example of how and where you have followed this command in everyday life. (About clothes? food? events? places?)
   b. Now repeat **a** for two of the other commands.
   c. Will **A** help you to follow these commands in geography? **Discuss** this with a partner. (No need to write!)

2. *Ironbridge benefits from being on the River Severn.* **Discuss** the statement in italics.

   Help for question 2
   - **Write your opening sentence.**
     It could say that Ironbridge lies on the River Severn.
   - **Describe how the river benefits Ironbridge.**
     For example tourists come to see the bridge. Would it exist without the river? Does the river make the place attractive? (See photo **B** on page 13.) Anything else?
   - **Describe any negative impacts of the river.**
     Floods? Start with *But …* or *However …*
   - **Write your closing sentence.**
     Say whether the benefits outweigh the negative impacts. You could begin with *On balance …*

3. When the water level in the River Severn rises, Ironbridge is at risk of flooding. In 2004 they began to use temporary flood barriers when floods were on the way. (See photo **E** on page 17). Look at this table for Ironbridge:

Year of flood	How far the river rose above average	Number of properties flooded
2000	7.04 metres	21
2008	6.05 metres	0
2014	6.02 metres	0

   Using the data in the table, **evaluate** whether the flood barriers are successful at protecting Ironbridge.

   Help for question 3
   - **Write your opening sentence.**
     It could say that Ironbridge is at risk of flooding, so temporary flood barriers were introduced in 2004.
   - **Give evidence that the flood barriers work.**
   - **Can you be sure they will always work?**
     (Could the river rise higher?) Start with *However …*
   - **Write your closing sentence.**
     Say whether the barriers are worth having, in your view. You could begin with *On balance …*

4. **Examine** how the natural resources of the Ironbridge Gorge influenced the rise of industry there. (Page 12!)

   Help for question 4
   - **Write your opening sentence.**
     It could name the natural resources found in the area. Include the river?
   - **Describe the role of each resource in the local industries.**
     What resources were used to make iron? What were the clays used for? What about the river?
   - **Write your closing sentence.**
     You could say that each resource played a key part in the rise of industry. You could begin with *To summarise …*

5. **Assess** how well the Ironbridge Gorge caters for tourists, in terms of parking areas. (The map on page 15 will help.)

   Help for question 5
   - **Write your opening sentence.**
     It could say that the Ironbridge Gorge gets thousands of tourists each year.
   - **State the number of parking areas there are – and where they are.**
     Does each museum have a parking area nearby? Do some have more than one?
   - **Write your closing sentence.**
     Say whether the Ironbridge Gorge appears to provide enough parking areas. You could begin with *To summarise …*

6. **To what extent** is the Ironbridge Gorge a suitable destination for a school trip from your school?
   (Note that it has lots of places to eat, and a Youth Hostel where school groups can stay. Use this information and what you know from earlier units to answer the question.)

   Help for question 6
   - **Write your opening sentence(s).**
     It could say where the Ironbridge Gorge is.
   - **Give reasons why the Ironbridge Gorge is a good place for a school trip.**
     Its history? Lots to look at? Food? Accommodation? Parking? Anything else?
   - **Are there any drawbacks?**
     For example is it a long way from your school? Start with *However, there is a drawback /there are drawbacks …*
   - **Write your closing sentence.**
     Give your final opinion about the Ironbridge Gorge as a destination. Begin with *On balance …* or *In conclusion …*

1 Geography …and you

How much have you learned about studying geography? Let's see.

check

1. Panel **A** has information related to your geography kit.
   a. Define the term *aerial photo*.
   b. The first aerial photo was taken in 1858. How did the photographer get into the air?
   c. Google Earth could not have existed before 1990. Explain why.
   d. i What does *GIS* stand for?
      ii Explain why GIS is so useful.
      iii Give two differences between GIS and paper maps.

2. **B** shows a wildfire in California, USA, in 2018. It killed 86 people and destroyed most of the town of Paradise. The image was captured by a Landsat satellite which orbits Earth over 14 times a day, 705 km above us.
   a. Define the term *wildfire*.
   b. What is the cloud of grey and white stuff?
   c. Suggest two reasons why it's better to use a satellite to track a wildfire than to take aerial photos from a plane or drone.
   d. Suggest one way in which a satellite image like this one could help firefighters.

3. Change, and the impact of change, are big themes in geography. Here are two changes which involve Earth:
   – a wildfire
   – global warming (temperatures rising around the world)
   a. Which of the two changes is more rapid?
   b. Which is likely to have a greater impact on humans and wildlife in the long term? Explain your answer.

4. Your teacher asks you to explore Tanzania, the country shown in map **C**, from your desk. So it's time to get nosy! The first question you could ask is: *Where is Tanzania?*
   a. Make up eight other questions about the geography of Tanzania, covering different aspects. (Pages 6 and 10?)
   b. For each, add a letter to say which strand(s) of geography it covers: physical (P), human (H), or environmental (E).
   c. Beside each question, say where you'll go to find the answer. An atlas? Google Earth? Google images? Another resource?

5. **D** gives data for the population of the Ironbridge Gorge.
   a. Define *population*. (Glossary?)
   b. Calculate the rise in population from 1801 to 1860.
   c. From what you know about the history of the Ironbridge Gorge, suggest a reason for:
      i the overall rise in population from 1801 to 1860.
      ii the overall fall in population from 1860 to today

6. *Geography helps us to understand the world.*
   Discuss the statement in italics.

What?	Date (CE)
oldest known world map	5th century
first balloon flight	1783
first camera	1816
first flight by plane	1903
first satellite launch	1957
first mobile phone call	1973
internet (world wide web) invented	1990

Year	Population (rough figure)
1801	4760
1860	9500
today	4000

2 Maps and mapping

2.1 Mapping through the ages

Maps are the top resource in your geography kit. So how has mapping changed over the years? Find out here.

What's a map?

A **map** is a drawing of a place, as from above. How accurate it is depends on:

- how much the map maker knows about the place
- how carefully it has been measured.

Over the centuries, we humans have learned more and more about places, thanks to explorers, and traders, and travellers of all kinds – and now thanks to satellites. Instruments for measuring have also got better and better.

So maps have changed a lot.

Two maps of the world

Compare these two maps of the world, drawn about 300 years apart.

▲ *Explorers were map makers too – like the Italian explorer Vespucci (1454 – 1512). Here he's working out where he is, using the stars.*

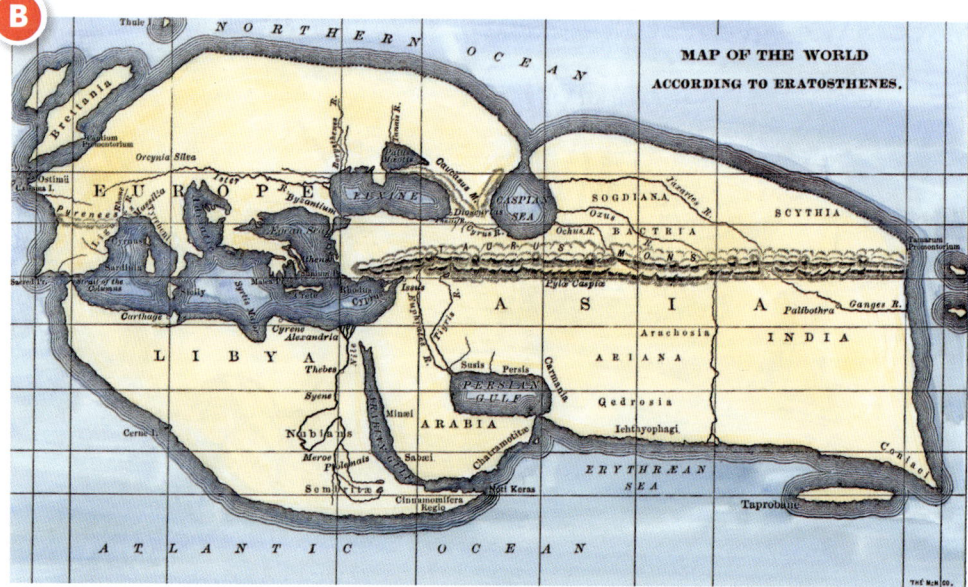

The world in 500 BCE, as mapped by the Greek geographer Hecataeus. At that time people thought Earth was flat. You could sail off the edge!

The world around 194 BCE, as mapped by the Greek scholar Eratosthenes. By then Greek astronomers had proved that Earth was not flat, but a sphere. Eratosthenes drew his map with the help of earlier maps, and travel books. (He was not an explorer.)

Note that map making or **cartography** went on in many places, not just Greece! Now turn to pages 140 – 141 to see the world as we know it today.

Measuring

For thousands of years, people relied on the sun, moon, stars and planets to work out where they were. For example, by measuring the sun's height in the sky at noon you can work out how far north or south you are.

Explorers made their own maps. So they needed to know about astronomy, and how to take measurements and do calculations, and how to look up the tables of data that astronomers provided.

For measuring, they used instruments like the **astrolabe** on the right.

▼ *The astrolabe, invented around 200 BCE. It lets you measure the angle between the sun (or other star) and the horizon. Then you look up tables to see how far north or south you are. Vespucci (above) is using one.*

The compass

The **compass** was invented in China. From around 1200 CE, it was widely used for finding direction. The photo on the right shows a compass that Captain Cook may have used.

The traditional compass has a needle that swings to line itself up with Earth's magnetic field. So it points north / south.

Compasses are still used today. Ships and planes carry them as back-up. The design and materials may have changed, but the purpose is still the same: to show direction.

Mapping today ... with help from satellites

Today, map making has changed hugely, thanks to **satellites**. And maps have become more and more accurate.

As you saw on page 9, hundreds of satellites are orbiting Earth right now, capturing images. These images help us to make maps of even the most remote places.

Other satellites help us to work out where on Earth we are, without stars or compasses. The **GPS** or **Global Positioning System** is an example.

GPS has over 30 satellites orbiting Earth. They send out radio waves. Down on Earth, a GPS receiver detects the radio waves and uses them to calculate its exact position on Earth. Perfect for drawing maps!

Mobile phones use GPS. So does the **sat nav** (satellite navigation) system in cars. It's hard to get lost!

▲ It's thought that this compass was used by the explorer Captain Cook on at least one of his three great sea voyages.

▲ Sat nav is a GPS receiver plus road maps. Type in the postcode for where you want to go, and it will show you the best route. The blue triangle here represents the car. Its position is continually updated.

Your turn

1. Define: **a** cartography **b** cartographer (Glossary?)
2. **a** Compare maps **A** and **B**, and identify:
   - **i** two similarities between them **ii** two differences
   b About how long ago was map **B** drawn?
3. Map **B** shows more land, and has more detail, than map **A**. (Look at the River Nile, for example.) Suggest a reason.
4. Both **A** and **B** show a sea between Libya and Europe.
   a Name this sea. The map on page 106 may help.
   b The maps show quite a lot of detail around this sea, compared with other areas. Suggest a reason. (Hint: where is Greece?)
5. *Libya* is the name the Greeks used for the part of Africa they knew. So **B** shows parts of three continents.
   a Four other continents are not shown on map **B**.
   - **i** Name them. (Page 100 may help.)
   - **ii** Suggest a reason why they are not shown.
   b Great Britain is on map **B**. What is it called?
6. Compare map **B** with our world map on pages 140 – 141. Apart from showing all the continents, list any three other ways in which our world map differs from **B**.
7. **a** What does *GPS* stand for?
   b Describe how GPS works.
   c Explain why GPS can help us to draw accurate maps.
8. This diagram shows the main directions marked on a compass.
   a What do the four letters stand for?
   b Make a mental image of the diagram. Then cover it and draw your own copy. (Don't mix up W and E! They spell **WE**.)
9. Turn to the world map on pages 140 – 141. As usual with maps, north is at the top. Name the country that lies:
   a just north of Mexico   **b** just south of Mongolia
   c just west of Argentina   **d** just east of Iraq

2.2 Plans and scale

A plan is a map of a small area. So let's start with plans, to help you understand maps and scales.

First, a photo

Violet lives in Warkworth, a village in Northumberland. Here's a photo of her room, taken from the doorway.

It looks big – but cameras can distort. To say how big, you'd need to measure it.

▲ Violet … 653, 654, 655. A new PB.

▲ Violet's room, looking tidy. A new PB.

Next, a plan

This is a **plan** of Violet's room – a drawing of what you'd see looking down at it from above.

A plan is really a map of a small area – for example a room, or a house, or your school.

The scale

The plan is a view of the room, but shrunk. 1 cm on the plan stands for 25 cm in the room. That is the **scale** of the plan.

You can show scale in three ways:

1. In words: 1 cm to 25 cm
2. As a ratio: 1 : 25 (say it as *1 to 25*)
3. As a line divided into centimetres and then labelled, like this:

   0   25   50   75   100 cm

The scale is always marked on a plan, so that people can tell the size in real life.

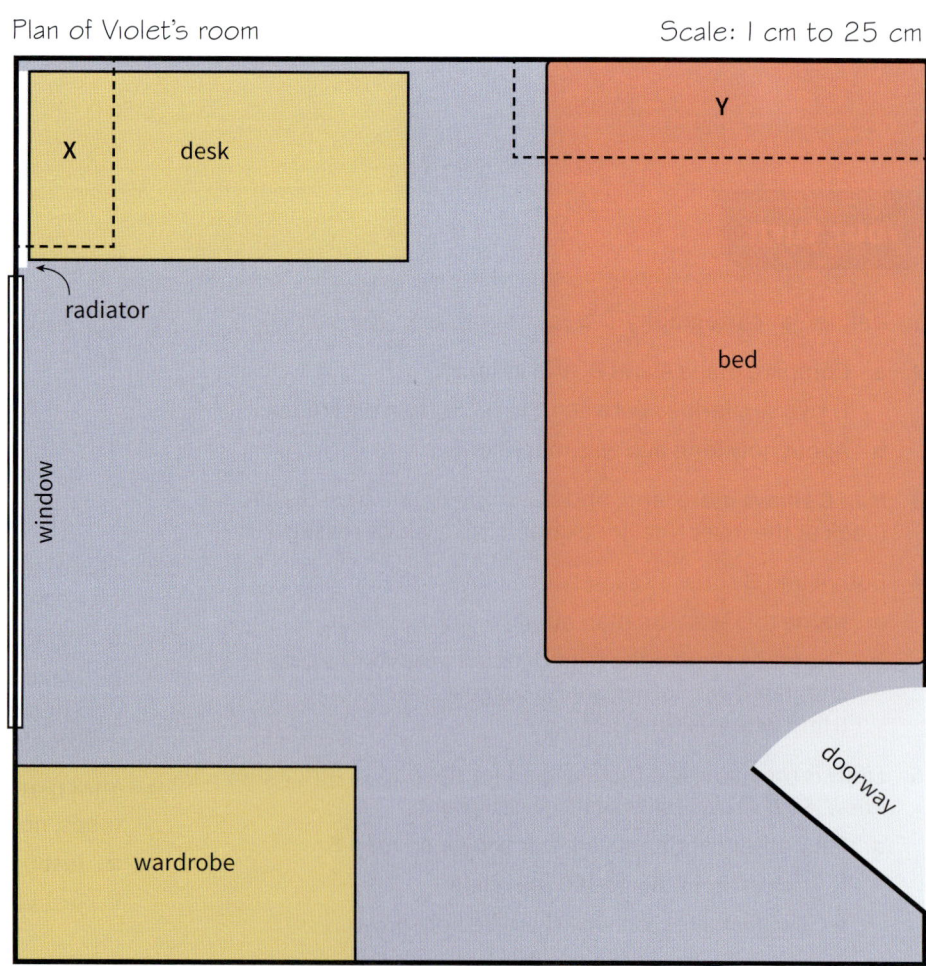

Maps and mapping

Working out scale

This is the plan of a table in Violet's kitchen. The table is 8 cm long on the plan. It is 160 cm long in real life.

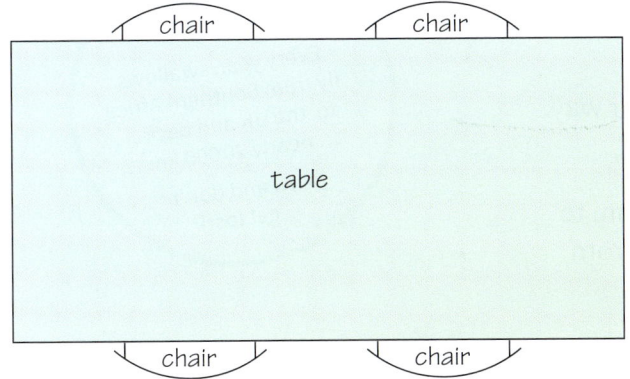

- 8 cm on the plan represents 160 cm in real life.
- So 1 cm on the plan represents 20 cm in real life.
- So you can write the scale as:
  1 : 20   or   1 cm to 20 cm   or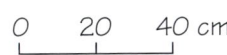

Be careful with units!

Look at this scale.

0 2 4 6 8 10 12 m

Here 1 cm represents 2 metres.
You can write this as **1 : 200**.

The 2 metres has been changed to centimetres. That's because *you must use the same units on each side of the* **:**.

1 : 200 means **1 cm to 200 cm** or **1 cm to 2 m**.

And remember …

100 centimetres (cm) = 1 metre (m)
1000 metres (m) = 1 kilometre (km)

Your turn

You will need a ruler for these questions.

1. Look at the plan of Violet's room. What do **X** and **Y** represent? (Check the photo!)

2. On a plan of a different room, one wall is shown like this:
   _____
   This plan uses 1 cm to represent 60 cm in real life.
   So the scale is 1 : 60. How long is the wall in real life?

3. Below are walls from another plan. The scale is 1 : 50. Calculate the length of each wall in real life.
   a _____
   b _____

4. Using a scale of 1 cm to 20 cm, draw a line to represent:
   a 40 cm   b 80 cm   c 2 metres
   Write the scale beside your lines.

5. If the scale is 1 : 300, what length does each line represent? Give your answer in metres.
   a _____
   b _____
   c _____

6. A kilometre is 1000 metres. Draw a line to represent 1 kilometre using each of these scales:
   1 cm to 1 km   1 cm to 200 m   1 cm to 100 m
   Write the scale beside each line, in any form you wish.

7. Make a chart like this and fill it in for Violet's room.

Violet's room	On the plan	In real life
How wide is it? Measure the wall behind the desk.		
How long?		
How long is the bed?		
How wide is the window?		
How wide is the doorway?		

8. Here is a new chest of drawers for Violet's room.

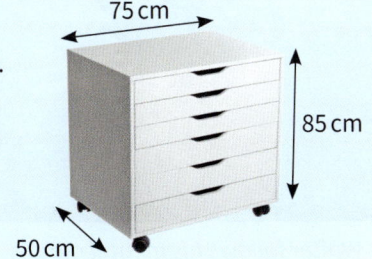

   a To draw a plan of it, which surface will you use?
      the top   the side   the front
   b Draw a plan of it, to the scale used for Violet's room.
   c Will the chest of drawers fit through the doorway?

9. Find three things in the photo of Violet's room that are not shown on the plan. *Why* are they not shown?

10. How would you draw a plan for your room? Write a set of instructions to follow.

25

2.3 The maps in your head

You are a map maker! You have made lots of maps in your head. Here you'll think about them – and get a chance to sketch one.

Mental maps

A **mental map** is a map that you make in your head.

It is really a sequence of images, like a movie. It helps you find your way.

You have lots of mental maps. You use them without even thinking.

You have one of your home, which helps you get from your bedroom to the bathroom or kitchen, even in the dark. And one of your route from home to school, and to shops you like.

Did you know?
- Every year, swallows fly from Southern Africa to the UK and back … … nearly 20 000 km …
- … and don't get lost!

Sketching a mental map

It's fun to 'see' your mental map in your head, then draw it on paper.

You end up with a rough map or **sketch map**.

Look at the sketch map on the right. It was drawn by Walter, Violet's cousin, who lives in Liverpool. It shows his local area.

Would you find it easy to follow?

▲ Walter follows his mental map to the post office, with a mystery parcel for his cousin Violet.

key
- mainly houses
- roads
- park
- trees
- graves

A MY LOCAL AREA

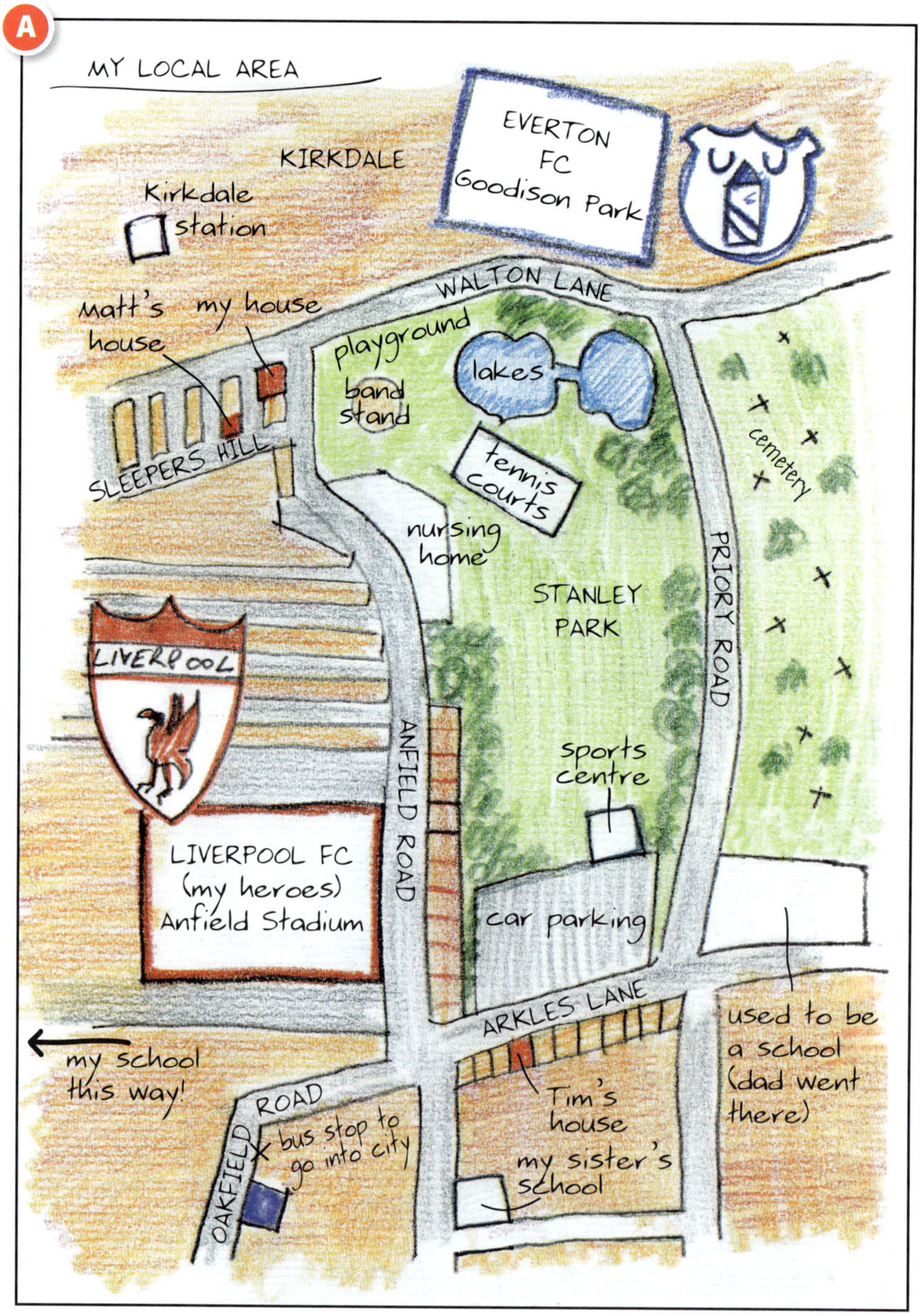

Maps and mapping

Your own mental maps

You have mental maps of your home, and your local area.

But that's not all. You have mental maps of other places you visit, and places you see on TV, and in computer games. You have mental maps of the UK, and the world.

On the right is Walter's sketch map of Britain, drawn from his mental map. What do you think of it?

Your mental maps are gappy

Our mental maps show things that are important to us. Such as paths we use, shops we like, places we have fun.

But they leave out lots of things. Some have big big gaps. Some are quite wrong, and can get you lost.

You can make them better

You can make your mental maps better and better. The secret is: *Look around. Keep your eyes open. Observe!*

It's fun to build up your mental maps, and fill in places. It's like a game.

The better your mental maps are, the better your grasp of your world.

▲ How Walter 'sees' Britain.

What if…
…your mental maps were almost blank?

Your turn

1. What is a *mental map*?

2. Think about *your* mental maps. See how many you can list. For example, do you have one of your route to school?

3. Look at Walter's sketch map **A**, on page 26.
   a. List the things he marked on it.
   b. Beside each, suggest a reason why he picked it.

4. Is Walter's sketch map easy to follow? Let's see! Give directions to get by road:
   a. from Walter's front door, on Anfield Road, to Tim's house. You could start like this:
      • Go out front door and turn right.
      • Walk along ____ ____ until …
      • Then …
   b. from Tim's house to the bus stop into town
   c. from the corner of Walton Lane and Priory Road, to Anfield Stadium

5. a. Now, take a few minutes to picture the area around your school, in your head.
   b. Using your mental map, draw a sketch map of the area. You can colour it in if you like.
   c. Compare your sketch map with your partner's. Do both show the same things?
   d. Suggest a reason why people might have different mental maps of the same area.

6. Look at **B**, Walter's sketch map of Britain. Compare it with the atlas map on page 139. Is the shape roughly right? Are his towns and cities in the right places? Give him a score out of 10.

7. a. Over the next week, pay special attention to the area around school. Look around. Keep your eyes open. Note the names of streets and roads. Observe!
   b. Then update your map from **5b** with any new details from your mental map. For example, street names.

2.4 From an aerial photo to a map

Here you'll compare a photo, a sketch map, and maps drawn to scale.

First, the aerial photo
This photo shows Warkworth in Northumberland, where Violet lives. Walter likes staying here.

It's an **aerial photo** – taken from the air. Look at the loop of the river, and the Norman castle.

Next, the sketch map
Below is a sketch map of the same place, which Walter started. He drew it from the photo. (You'll do that too.)

Note that his sketch map has:
- a title, a frame, and a key
- some labels and annotations (notes)
- just enough detail to show the shape and layout of Warkworth. (Not each building and tree!)

A
- all the open land outside the loop of the river is farmland
- bridges
- farmland
- River Coquet
- remains of Norman castle
- Violet's house

▲ Warkworth from the air.

B Warkworth, where my cousin Violet lives. (Not to scale.)

- There is farmland all around the village.
- These houses are tucked into the loop of the river.
- bridges
- castle
- The castle was built over 900 years ago by Normans, but rebuilt later. Quite a lot of it is in ruins.

Key
- river
- trees
- homes and gardens
- road
- farm land
- open green areas

Warkworth

28

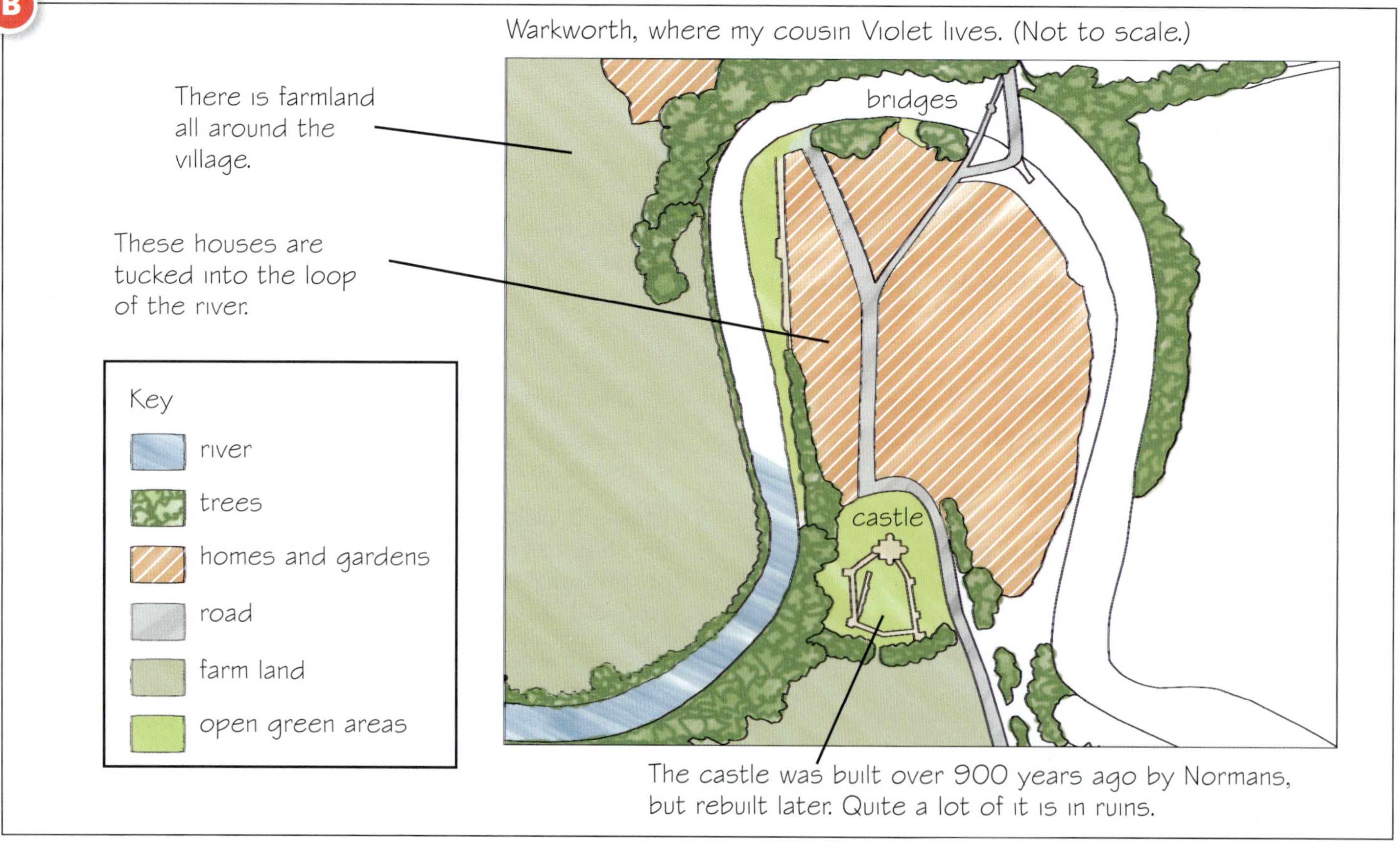

Now, a map drawn to scale

Look at this map of Warkworth.

It is not a sketch map. It is an accurate map, drawn to scale. (Look for the scale below.)

It uses symbols to show things. They are given in the key.

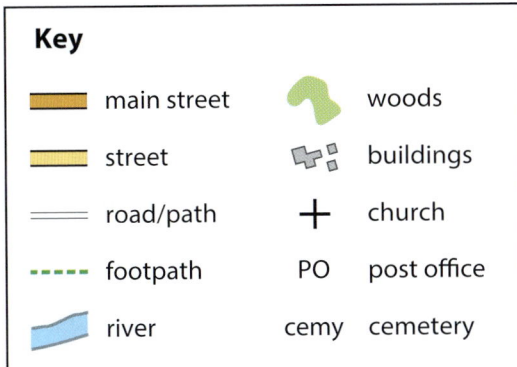

The same map with a grid

Here is the same map again. But this time, **grid lines** have been added.

The grid lines divide the map into squares.

The columns and rows of squares have been labelled (A, B ... and 1, 2 ...).

The post office is in square C3.

The cemetery is in square C4.

You always give the letter for the column first.

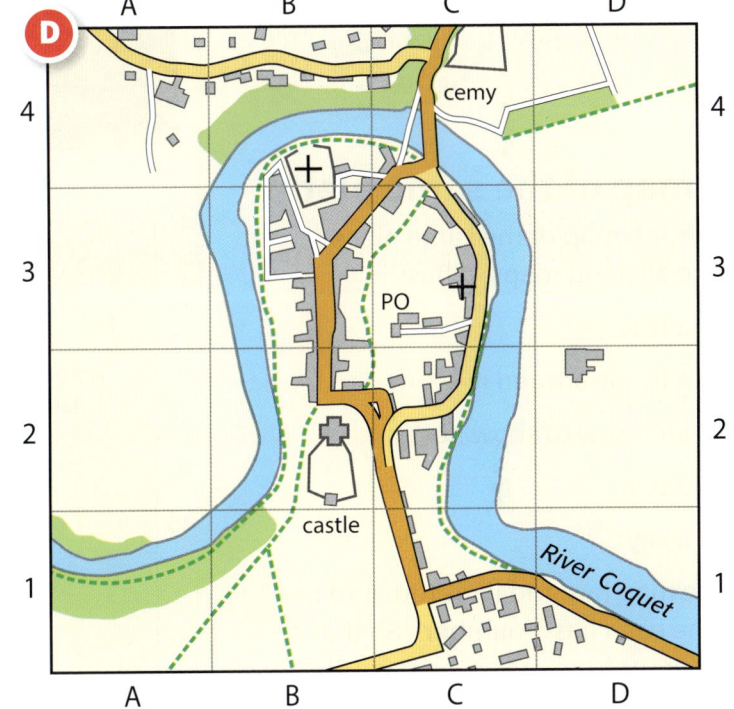

Your turn

1 Draw a sketch map (like the one Walter started) for photo **A** on page 28. Keep it simple. And don't forget:
   – a title, a frame, and a key
   – labels and annotations

2 Now swop sketch maps with your partner.
   a First, agree on a fair way to score the maps. For example you could give a mark out of 10 for the shape, 1 mark for each correct label, and so on. Write a list.
   b Then give each other's maps a score.

3 Look at map **C** above. In which ways is it:
   a like your sketch map?   b different from it?

4 a Describe where the castle is, on map **C**.
   b Did you find the task in **a** difficult? If yes, explain why.

5 On map **D**, the post office is in square C3.
   a Give the square for: **i** the castle **ii** the bridges
   b What is in: **i** square B4? **ii** square C1?

6 Are the grid lines in map **D** helpful, or a nuisance? Decide, and justify your answer.

2.5 Using grid references

In this unit you will learn how to find places on a map, using grid lines with numbers on.

A photo

This aerial photo shows part of the River Mole valley in Surrey.

In the top right is the village of Mickleham.

Walter went fishing in the Mole when he visited his other cousin Kim. (Fish fled.)

Did you know?
- The first ever aerial photo was taken in 1858, over Paris, from a hot air balloon.

A map of the same place

This is a map of the same place. Like all good maps, it has:

- a title
- a frame around it
- an arrow to show north
- a scale
- a key.

The map has grid lines too. And this time each has a number. (So that's different from map **D** on page 29.)

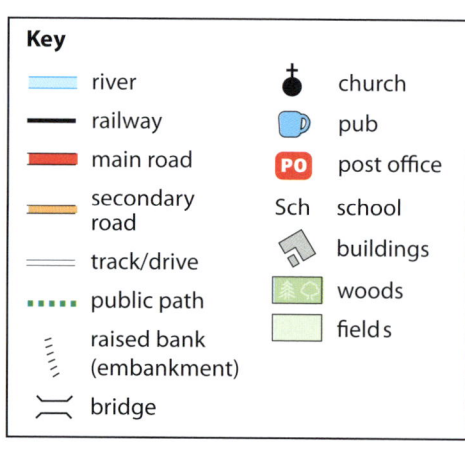

The River Mole valley near Mickleham

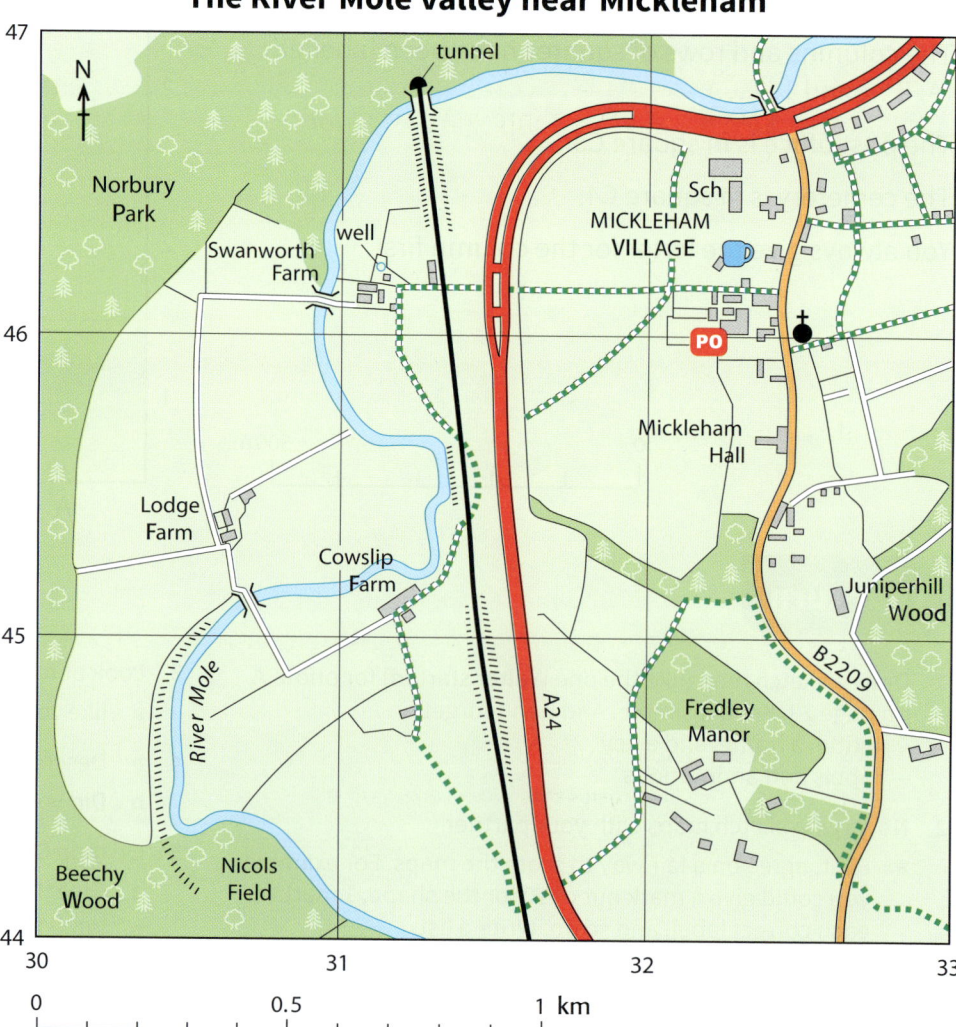

Maps and mapping

Four-figure grid references

Grid references are made from the numbers on the grid lines.
They help you to find a place quickly.
The school is at grid reference 3246. Fredley Manor is at 3244. Look:

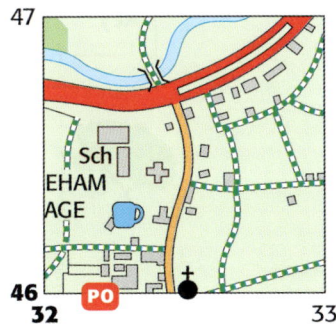

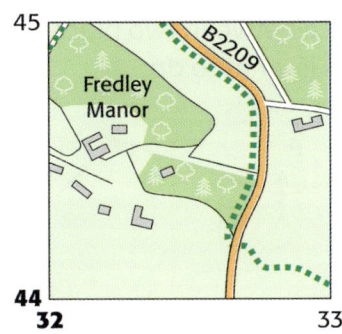

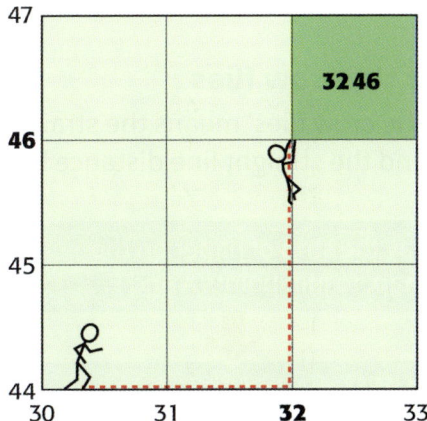

To find the school, go to the square where grid lines 32 and 46 meet in the lower left corner. Then look for the label 'Sch'.

To find Fredley Manor, go to the square where grid lines 32 and 44 meet in the lower left corner. Then look for the manor.

A grid reference always gives the number along the bottom first. This drawing shows how to find square 3246. **Walk before you climb!**

The grid references above are called **four-figure**. Why?

Six-figure grid references

Square 3246 has a school *and* a church. You can add two extra numbers to say *where* each is in the square. Like this:

- Divide the sides of the square into ten parts in your head, as shown on the right.
- Count how many parts you must walk along to reach the building, and how many parts you must climb.

For the school you go <u>3</u> parts along and <u>5</u> parts up.
So its six-figure grid reference is 32<u>3</u>46<u>5</u>.
The one for the church is 32<u>5</u>46<u>0</u>. Do you agree?

We usually show all six of the numbers in black.

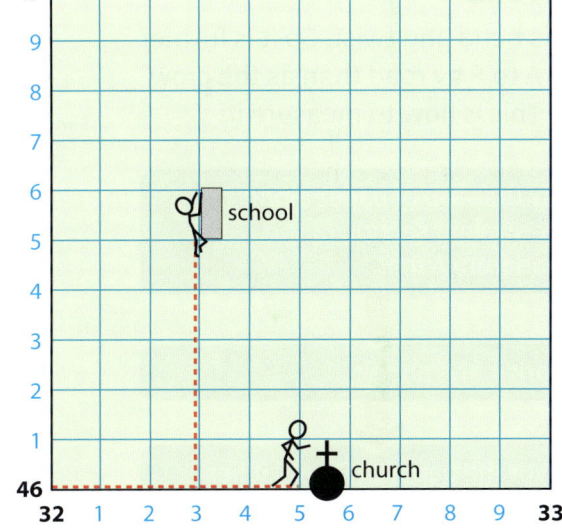

Your turn

1. Look at the map on page 30. Name two things you can see in the square with this grid reference:
   - **a** 3245   **b** 3246   **c** 3046   **d** 3144
2. Give a four-figure grid reference for:
   - **a** Lodge Farm   **b** Cowslip Farm   **c** Nicols Field
3. Identify the feature at this grid reference on the map:
   - **a** 312468   **b** 309449   **c** 309461
4. Give a six-figure grid reference for:
   - **a** Mickleham Hall   **b** the post office   **c** the pub

5. You can't *see* the river on the photo. How can you tell where it is? Explain.
6. Describe what you will see, if you stand at 313453 facing south. (With your back to the north!)
7. How far is it from Lodge Farm to Cowslip Farm, along the track? See if you can think of a way to measure it, using the scale. (Would thread or a strip of paper help?)
8. When are six-figure grid references more helpful than four-figure grid references? Explain your answer.

2.6 How far?

Here you'll learn how to find the distance between two places, on a map.
You will need a strip of paper with a straight edge.

1 As the crow flies

'As the crow flies' means the straight line distance between two places.
To find the straight line distance from A to F, this is what to do:

① Lay the strip of paper on the map, to join points A and F.

② Mark it at A and F.

③ Now lay the paper along the scale line to find the distance AF.

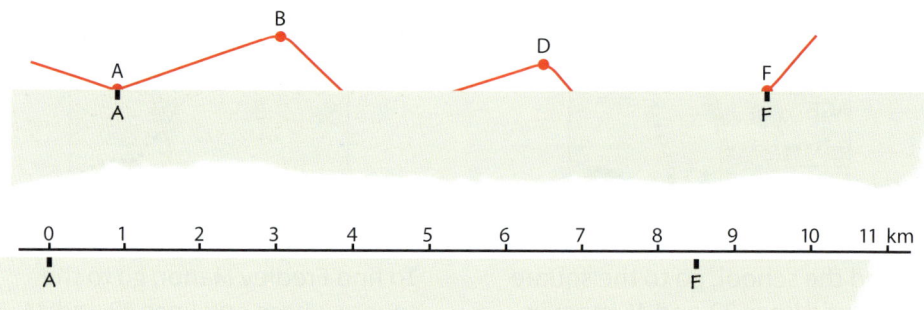

From A to F as the crow flies is 8.5 km.

2 By road

Roads bend and twist. So it is further from A to F by road than as the crow flies. This is how to measure it:

① Lay the strip of paper along the straight section of the road from A to B.

② Mark it at A and B.

③ Pivot the paper at B until it lies along the next straight section, B to C. Mark it at C.

④ Now pivot it at C so that it lies along the next straight section, C to D. Mark it at D.

⑤ Move along like this, section by section, until you reach F.

⑥ Place the paper along the scale line to find the distance AF.

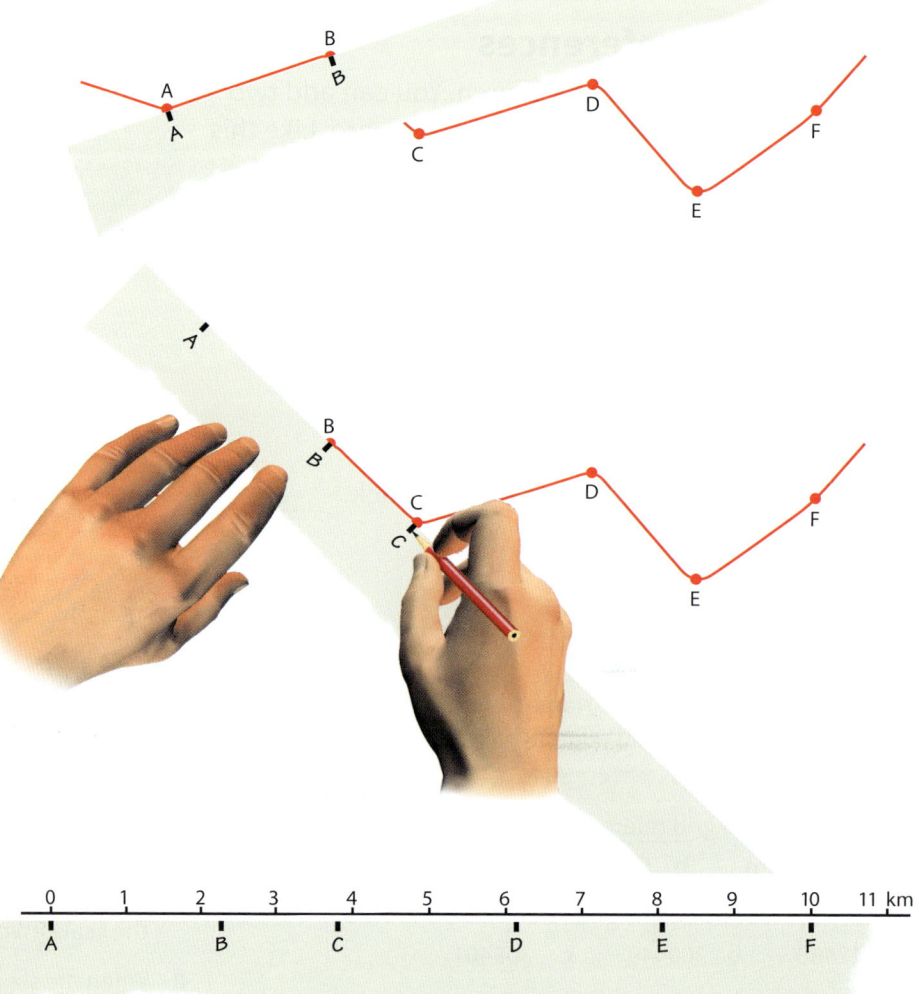

From A to F by road is 10 km.

Your turn

Maps and mapping

The photo and map on page 30 showed part of the River Mole valley in Surrey. This map shows more of the same area. (Are both maps at the same scale?)

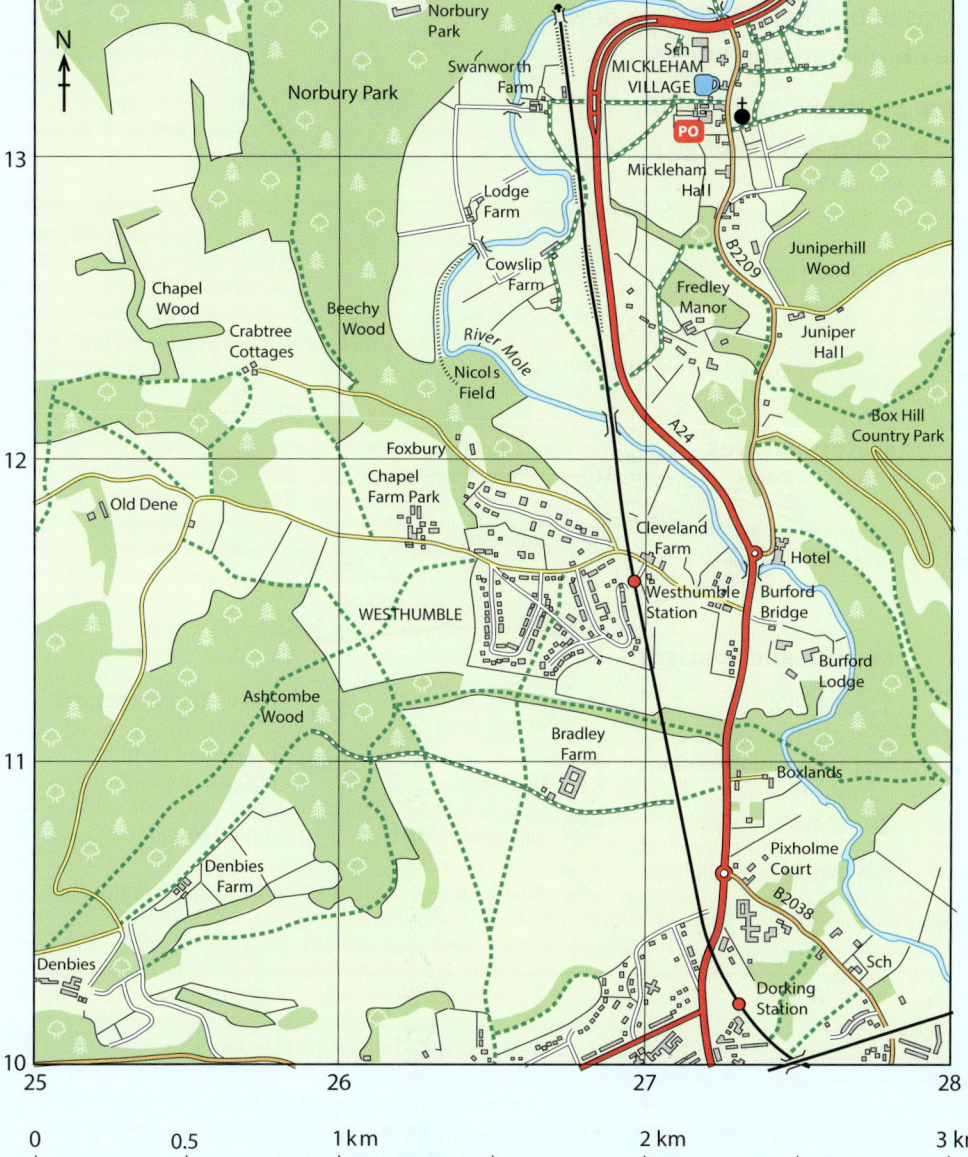

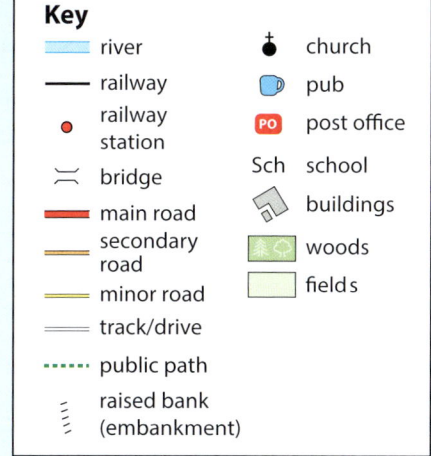

▲ St Michael's church, Mickleham.

▲ Juniper Hall.

1. How far is it from Mickleham church to Westhumble station, as the crow flies?

2. How far is it by rail from Westhumble station to Dorking station? Measure along the railway line.

3. About how far is it by road from Mickleham Hall (273129) to the hotel at 274117?

4. Walter arrived at Westhumble station, to visit his friend Joe. Joe had given him these directions:

   Walk from the station towards Cleveland Farm.
   At the junction with the minor road, turn left.
   At the next fork, take the road to the left, and walk for 0.7 km.
   Where does Joe live?

5. Every day, Kim's mother collects her from the school (Sch) at 276103 and drives her home by this route:

   From the school, go right on the B2038.
   At the roundabout, take the A24 north for 0.9km.
   Turn left onto the minor road, and continue for 0.5km.
   Now take the road to the right, and continue for 1.4km.
   Where does Kim live?

6. Juniper Hall and St Michael's church are shown above.
   a. Find them on the map, and give six-figure grid references for them.
   b. Your friend wants to walk from the church to Juniper Hall. Write instructions. Don't forget to give the distance!

2.7 Ordnance Survey maps

 Here you'll learn what OS maps are, and what they show, and how to use them.

What are OS maps?

Ordnance Survey maps or **OS maps** are maps of places, with lots of detail. They use symbols to show things. They have numbered grid lines.

The OS map opposite shows Warkworth (from page 28), and Amble. The key below has the symbols – and there's a larger key on page 138.

Did you know?
- The Ordnance Survey was set up in 1791, to map Britain for the army.

Key

Roads, paths and boundaries
- main road
- secondary road
- minor roads
- path
- footpath; bridleway
- edge of field

Railways
- railway track
- cutting; tunnel; embankment

Buildings
- building; important building
- places of worship (current and former) with tower; with spire, minaret or dome; without such additions

Abbreviations
- CH   club house
- PO   post office
- Sch school
- Cemy   cemetery
- LB Sta lifeboat station
- W; Spr well; spring

Vegetation
- Coniferous trees
- Non-coniferous trees
- Bracken, rough grassland, heath
- Marsh, reeds or saltings

Water features
Slopes, Cliff, Flat rock, Lighthouse, Sand Dunes, Mud, Shingle, Beacon

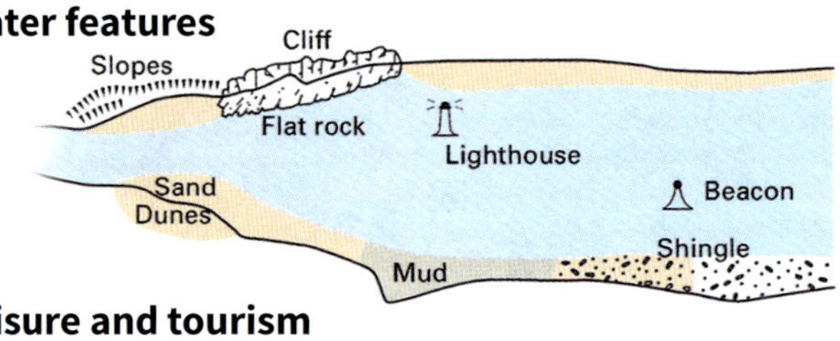

Leisure and tourism
- P parking
- information centre
- public phone
- public convenience (toilet)
- golf course or golf links
- boat trips
- slipway for boat
- picnic site
- run by English Heritage
- *Name* used for old and ancient sites
- nature reserve
- other tourist feature

Your turn

1. Look at the OS map. Name the river that flows through Warkworth. Which sea does it flow to? (Page 139?)
2. Find it on the map, and give its four-figure grid reference:
   a Northfield   b Gloster Hill   c North Pier
3. What is at this grid reference on the map?
   a 243045   b 277041   c 247057
   d 243065   e 236058   f 275049
4. 4 cm on this OS map represents ____ in real life. Complete!
5. The top of an OS map is always north. Look at the photo of Warkworth on page 28. Where is north on it?
6. Violet's house is marked on the photo on page 28. Find it on the photo. Then find it on the OS map, and write a six-figure grid reference for it.
7. Warkworth has a population of around 1600. Now look at Amble. Its population is about _____? Choose from these:
   a 1000   b 2000   c 6000   d 9300
   How did you decide?
8. How many of these are there in Amble?
   a schools   b places of worship   c cemeteries
9. Find one of these on the map and give a six-figure grid reference for it:
   a a post office   b a club house
   c a public phone box   d an *old* bridge
10. What clues are there on the map, that Warkworth and Amble get lots of visitors? Give as many as you can.
11. Write a travel blog about things for tourists to do in Warkworth and Amble. At least ten lines.

2.8 How high?

 In this unit you'll learn how height is shown on an OS map.

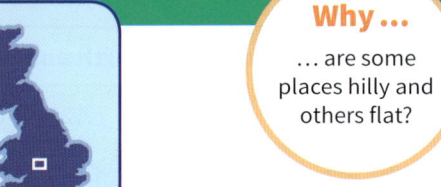

Why ... ... are some places hilly and others flat?

Land height around Temple Guiting

This aerial photo shows Temple Guiting, a small village in Gloucestershire where Violet's uncle lives. (Say *Temple Guy-ting*.)

▲ Measuring the height (altitude) of places above sea level used to be slow. Not now! He's using GPS, which tells him altitude as well as latitude and longitude. (Page 38.)

The OS map below shows the area around Temple Guiting – and tells you how high and hilly it is. The map shows height in two ways …

① **Contour lines** join all the places at the same height above sea level. The number on a line shows the height in metres. Here, the lines are every 10 m above sea level.

② **Spot heights** are small black numbers that give the exact height of a spot, in metres above sea level.

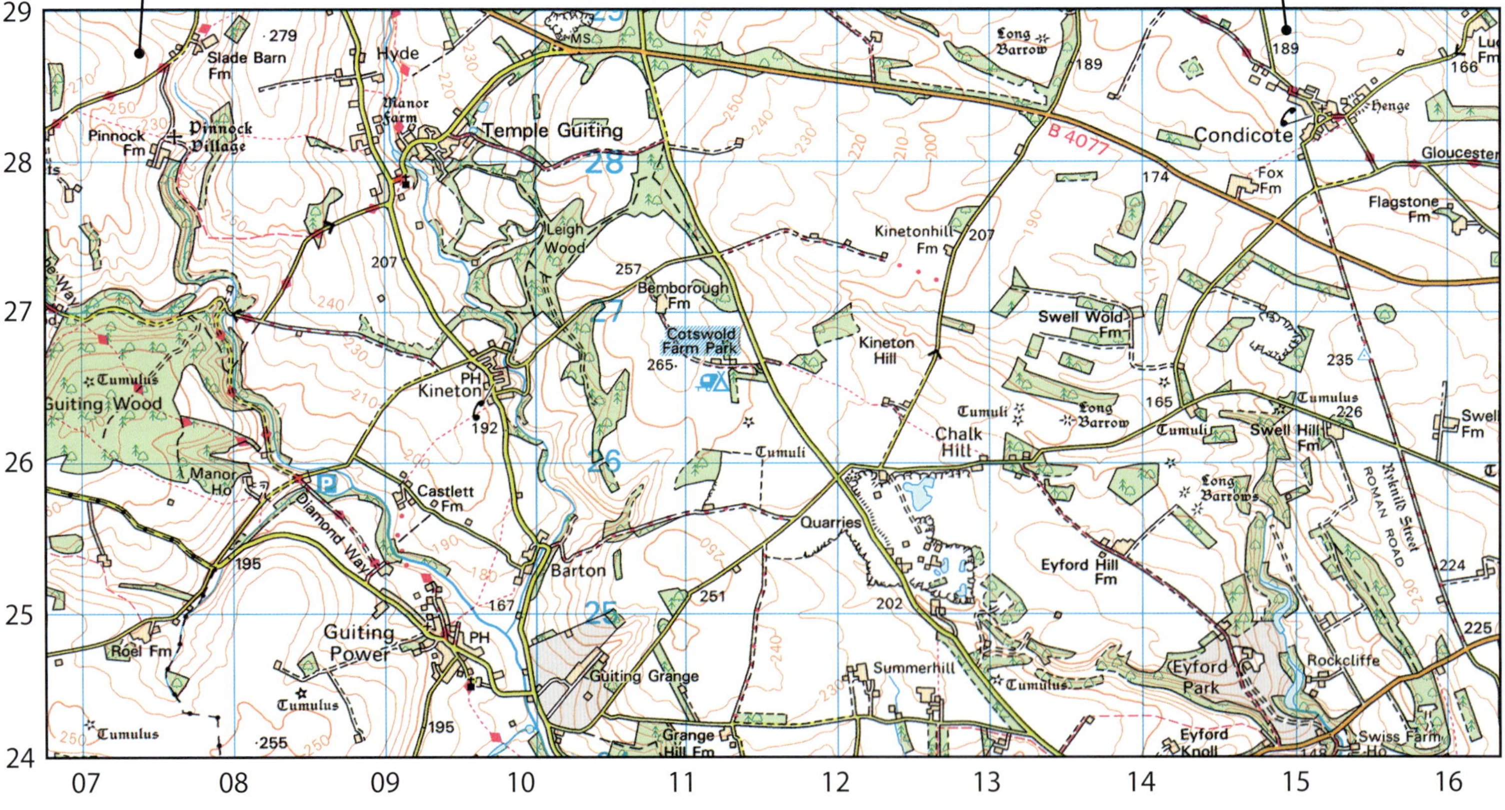

Scale 1: 50 000

© Crown copyright

36

More about contour lines

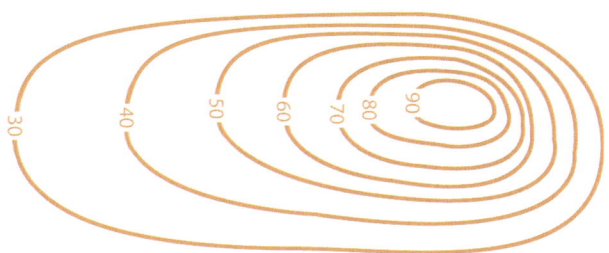

The contour lines are marked on this hill at 10 metre intervals. On a map, you see them from above …

… like this. They are close together where the slope is steep, and further apart where it is gentle.

Remember:
- where contour lines are very far apart, it means the ground is flat.
- where they are very close together, the ground slopes steeply.

Your turn

1. Define these terms: **a** contour line **b** spot height
2. Match the drawings to the contour lines. Start your answer like this: *A =*

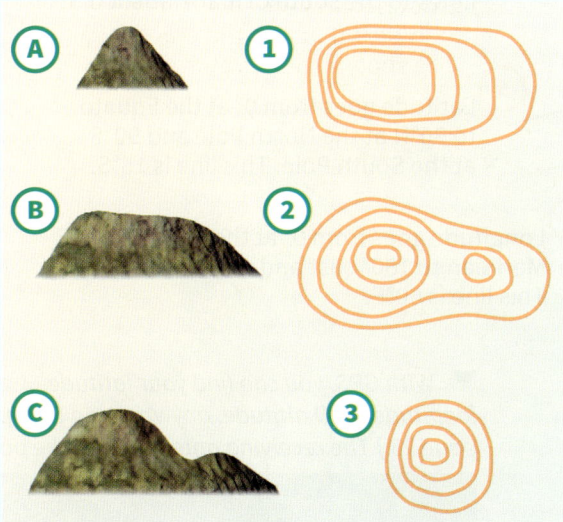

These questions are about the OS map on page 36.

3. The black line from panel **1** points to a contour line.
   a. What does the number on the contour line tell you?
   b. What height does the line to its left represent?
4. a. Give the grid reference for the square that contains the main part of Temple Guiting.
   b. How high is its church above sea level? Choose.
      i 200 m   **ii** 230 m   **iii** 250 m
      (Find a contour line to follow.)
5. a. The grid lines on the map are 2 cm apart. What distance does this represent on the ground? (Scale!)
   b. Castlett Farm lies south of the Temple Guiting church. About how far it is from the church as the crow flies?
   c. How high above sea level is Castlett Farm?
6. State whether you are going uphill or downhill, when you walk along the road:
   a. from the parking area in 0825 to Kineton in 0926
   b. from the crossroads in 1028 to the crossroads in 1328
   c. from Kineton to Bemborough Farm in 1027
7. a. In which square is the land flatter?
      i 0827   **ii** 1125
   b. Which square has steeper land?
      i 1025   **ii** 1228
   For each answer, justify your choice.
8. Kinetonhill Farm is about 2 km from Bemborough Farm. Which farm is at a greater **altitude** (height) above sea level? Use spot heights to help you decide.
9. Look at the river which flows beside Temple Guiting. It is the River Windrush. (It gave its name to the ship which brought people from the Caribbean to the UK in 1948.)
   a. About what height above sea level is the River Windrush:
      i at Temple Guiting?   **ii** at Barton?
   b. The top of an OS map is always north. Which way does the River Windrush flow – from north to south, or from south to north? Explain how you decided.
10. *It can be very important to know the altitude of a place. Suggest one example where altitude matters.*

2.9 Where on Earth?

Here you'll learn about the special grid lines we use to say where places are on Earth.

Why...
...doesn't Earth just roll around in space?

Did you know?
- Any line of longitude could be the Prime Meridian.
- Countries agreed to use the one that passes through Greenwich in London.

Grid lines around Earth

Earth is like a ball. So how do you say where you live, on a ball? You cover it with imaginary grid lines, and number them! This shows Earth with its grid lines. We call it a **globe**.

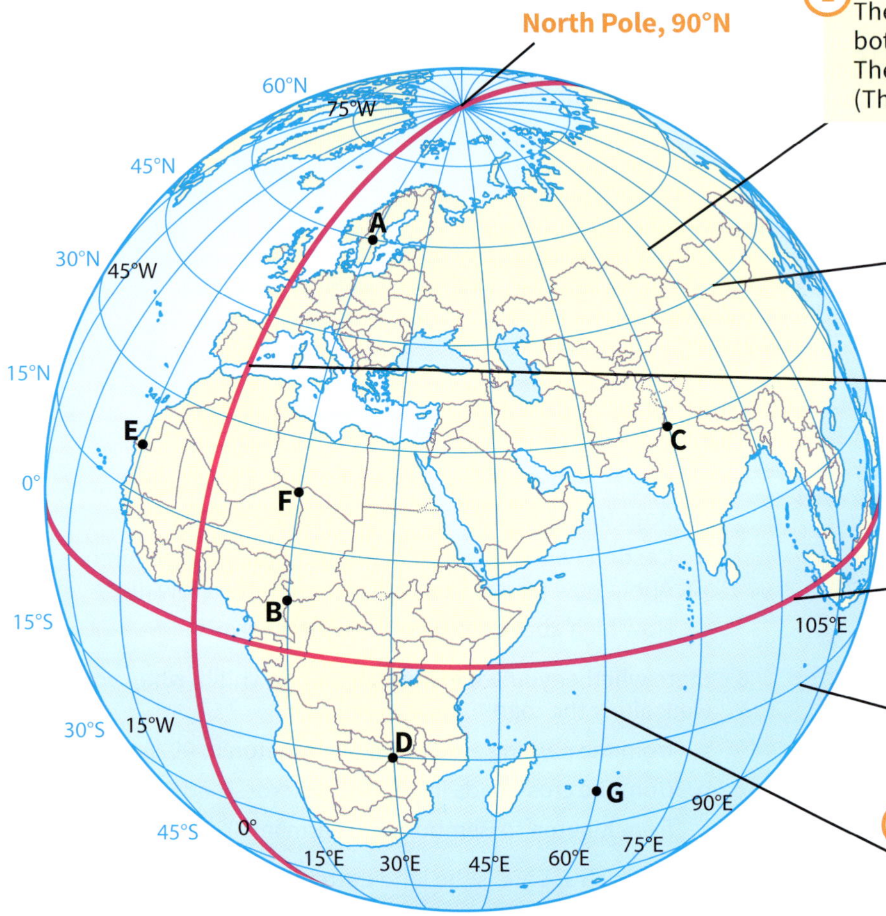

① The lines that circle Earth from top to bottom are called **lines of longitude**. They meet at the North and South Poles. (The South Pole is hidden on this drawing.)

② The lines that circle Earth from side to side are **lines of latitude**.

③ This special line is the **Prime Meridian**. Its longitude is taken as 0° (nought degrees). Lines to the east of it are labelled **E**. Lines to the west are labelled **W**.

④ This special line is the **Equator**. Its latitude is 0°. Lines to the north of it are labelled **N**. Lines to the south of it are labelled **S**.

⑤ Latitude goes from 0° at the Equator to 90°N at the North Pole and 90°S at the South Pole. This line is 15°S.

⑥ Longitude goes from 0° at the Prime Meridian to 180° east and west. This line is 60°E.

Here the grid lines are shown every 15°. But you could choose any interval. For example you could show them every 20° or 30°.

Using the lines to say where a place is

Coordinates

Look at place **A** on the globe. It is 60° north of the Equator, and 15° east of the Prime Meridian. So its **coordinates** are 60°N 15°E. You can give the position of any place on Earth using coordinates.

Degrees and minutes

Degrees are divided into **minutes**. 1 degree = 60 minutes, or 1° = 60'.

Look at **B**. It is halfway between 0° and 15° north of the Equator. So it is at 7° 30'N. Its coordinates are 7° 30'N 15° 00'E. Do you agree?

▼ With GPS, you can find your latitude, longitude, AND altitude, anywhere on Earth. (See page 23.) The receiving antenna is on the pole.

38

Maps and mapping

Showing Earth on a map

Earth is round. So how can we show it on a flat map? There are many ways to show it – but they all cause a little distortion. Here is one example:

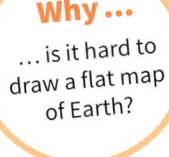

Why ...
... is it hard to draw a flat map of Earth?

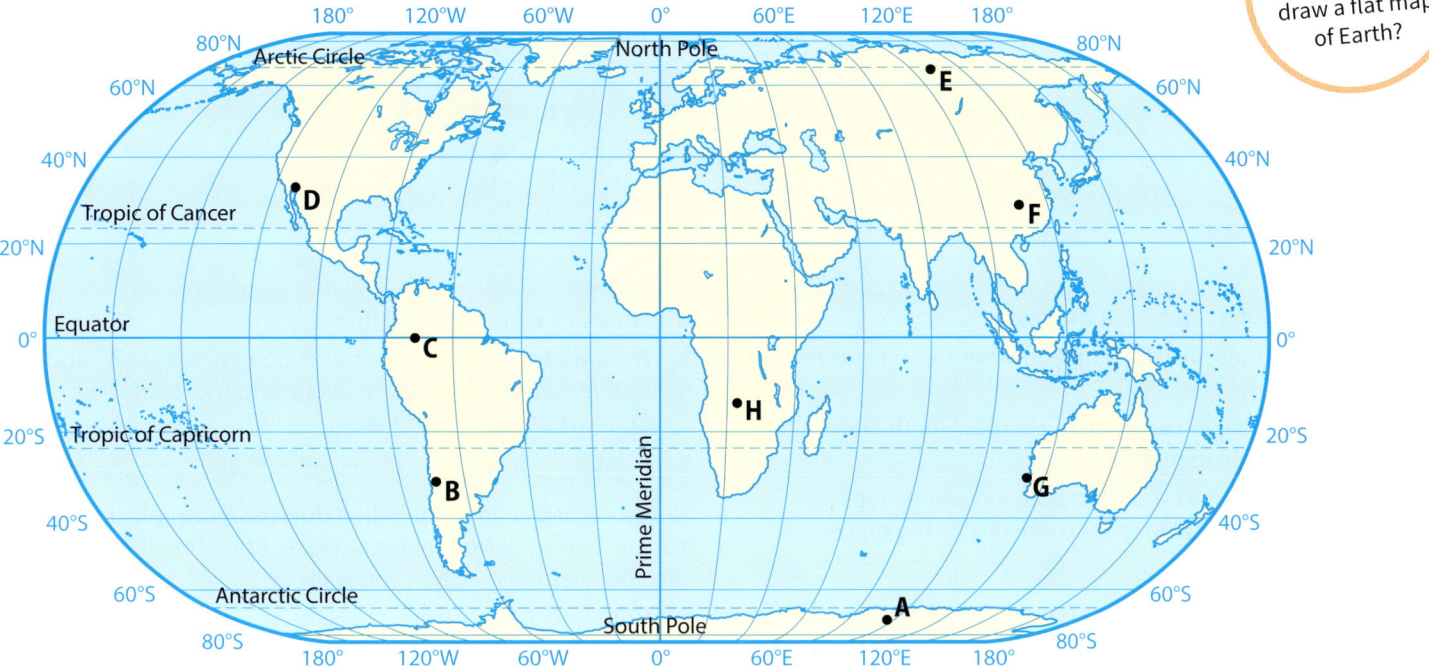

Here the land masses are the correct *sizes*, relative to each other. But their *shapes* are a little distorted. Compare Africa on the map and globe.

The five main lines of latitude

Look at these lines of latitude on the map above.

Equator	0°
Tropic of Cancer	23° 26'N
Tropic of Capricorn	23° 26'S
Arctic Circle	66° 33'N
Antarctic Circle	66° 33'S

They are all linked to the Sun's position in the sky, at certain times of year.

Look at the two tropic lines. The region between them is called the **tropics**. Look at the Arctic Circle. The region above it is called the **Arctic**. All the land below the Antarctic Circle is part of **Antarctica**.

Looking up places in an atlas

Suppose you want to find Paris.

- First, look for Paris in the index at the back of the atlas. You will find something like this:
  Paris, France **63** 48 52N 2 20E
- The first number is the page number to go to.
- The other numbers are the coordinates for Paris.
  (The ° and ' are often left out.)

Your turn

1. Define these terms: **a** Equator **b** Prime Meridian
2. Look at the globe on page 38.
   a Identify the place with these coordinates:
      i 30°N 75°E   **ii** 15°S 30°E   **iii** 60°N 15°E
   b Identify the place with these coordinates:
      i 22°30'N 15°E   **ii** 22°30'N 15°W   **iii** 7°30'N 15°00'E
   c Give the coordinates for place **G**.
3. For the map above, give the letters for these:
   a the place at this latitude: **i** 0°   **ii** 66° 33'N
   b two places in the tropics
   c a place in Antarctica
   d the place with these coordinates:
      i 34° 00'N 113° 30'W   **ii** 29° 50'S 116° 30'E
4. Now list all the countries the Equator passes through!

2 Maps and mapping

How much have you learned about maps and mapping? Let's see.

check

A

Scale 1:30

B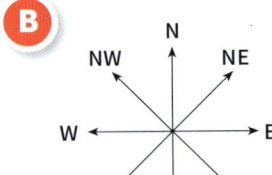

C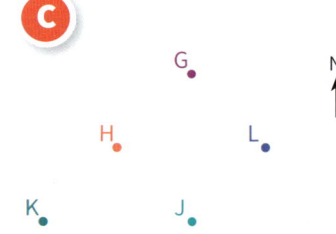

1 **A** shows the plan of a room, and the scale.
   a Use a ruler to measure the length and width of the room on the plan.
   b Now use the scale to calculate the actual length and width of the room in metres.
   c It is very important to show the scale on some maps – for example on a local map for tourists, like the one on page 15. Explain why.

2 Page 36 shows an OS map for the Temple Guiting area.
   a What does *OS* stand for?
   b OS maps show contour lines. What are contour lines?
   c The map has one feature which shows that the Roman army knew this area.
      i   Find the feature and state its name.
      ii Give a four-figure grid reference for one square which this feature passes through.
      iii Some spot heights are marked along this feature. State any one of these heights.
      iv Flagstone Farm lies east of the feature. Give a six-figure grid reference for the main building at Flagstone Farm.

3 You have to lead a group on a walk in the Temple Guiting area. They will ask you lots of questions about what's in the area. Which **one** of these will you bring with you?
   a a sketch map of a route, that you drew in advance
   b the OS map of the area
   c your mobile phone which has a map app that uses GPS, so you always know where you are
   Justify your choice, and explain why you didn't choose each of the others.

4 a What is a compass used for?
   b **B** shows the four main compass points (N, S, E, W) and four in-between points. NW stands for *North of West*. State what these stand for:
      i SW   ii NE   iii SE

5 Now look at **C**. The five labelled dots represent places. You have to walk from place to place. State the direction are you heading in, when you walk:
   a from G to J   b from J to G
   c from J to K   d from H to L
   e from L to G   f from H to K
   g from H to J   h from J to L
   Give your answer as a letter or letters. (For example N, SW.)

6 Look again at the OS map on page 36. You will walk:
   - from the crossroads in 1328 to the crossroads in 1028
   - and from there to the crossroads in 1125.
   Using a strip of paper and the map scale, calculate how far you will walk. (See the method shown on page 32.)

7 a Define these terms:
      i latitude   ii longitude   iii altitude
   b Here are the coordinates of three places, **X**, **Y** and **Z**:
      X, 15°S 60°W   **Y**, 30°S 130°W   **Z**, 42°N, 105°W
      For each place, identify the continent it's in.
      (Pages 39 and 100 will help.)

8 Sally loves driving. She's a taxi driver in the city in her spare time.
   a Outline how sat nav works.
   b Explain how sat nav helps Sally.

9 *Without maps, life would be much more difficult for us humans. To what extent do you agree with this statement? Write at least eight lines!*

3 About the UK

3.1 Your island home

This unit is about the group of islands where you live.

The British Isles

The map below shows the **relief** of the British Isles. Look at the mountainous areas, and flat areas. Look at the key.

▲ You'll find places like this in the British Isles …

▲ … and places like this.

Your turn

1 The map on page 42 shows the relief of the British Isles.
   a What is *relief*? (Glossary?)
   b The map shows lots of islands. State the name of:
      i   the largest island   ii   the second-largest one
   c On the largest island, where are:
      i   the highest mountains?   ii   the flattest land?
      Use terms like *north, south, east, west* in your answer.
   d Identify roughly where you live on the map, and describe the relief in that area. (Page 139 may help.)

2 Now look at the map below.
   a What does the orange colour on the map indicate?
   b Using the maps at the back of the book to help you, Identify the places and features marked **a – s**.

Key
a–j upland areas
k–n islands
o   a country
p–s sea areas

3 There are thousands of rivers in the UK. (Only the longest ones are shown on the relief map.) Using page 139 to help you, identify rivers **a – g** from these clues.
   a It's the longest river in the UK. It rises in Wales.
   b This one flows by the Houses of Parliament.
   c Stoke-on-Trent sits on this river.
   d Newcastle sits on this one.
   e This one runs along part of the border between England and Scotland.
   f Did Aberdeen get part of its name from this?
   g This one flows to the Wash, on the North Sea.

4 Photos **X** and **Y** above were taken at points **1** and **2** on this little map.
   a Which photo was taken at **1**? Explain your choice.
   b Now compare the two places shown in photos **X** and **Y**. (Say what – if anything – is similar about them, and what is different.) Write at least four sentences in your answer.
   c Name one job people might do, in each place.
   d Which of the two places would you rather spend a day in? Give reasons for your choice.

5 The British Isles have long coastlines which wiggle a lot. Name at least three jobs that depend on living at the coast.

6 a You live on an island. Is that a good thing? List any advantages you can think of, of living on an island.
   b Now list any disadvantages you can think of.
   c Which win, the advantages or the disadvantages? Justify your answer.

7 Write a paragraph saying where on Earth the British Isles are located. Pages 140 — 141 will help.
   You must include these terms in your paragraph:
   *Equator   Atlantic Ocean   continent   islands
   Europe   France   Arctic Circle   North Sea*

3.2 It's a jigsaw!

In this unit you'll see how we humans have carved up the British Isles.

Building borders

20 000 years ago there were no borders in these islands – because nobody was living here.

But over time, different tribes arrived. They fought over things like land, trade, and religion. And in the end …

… borders were built between different areas. We still have borders, and many other boundaries, today.

Two countries

Today, the British Isles is divided into two **countries**: the United Kingdom (UK) and the Republic of Ireland. The UK is in green on map **A** below. London is its capital.

The UK is in turn made up of four **nations**: England, Wales, Scotland, and Northern Ireland.

A

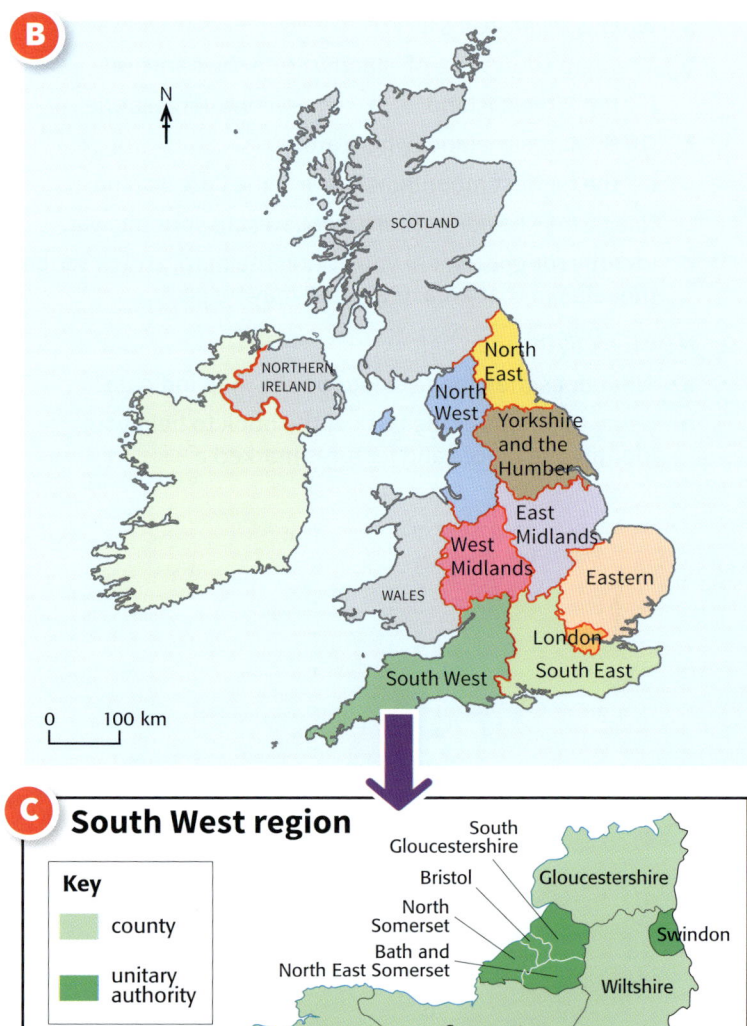

B

C **South West region**

Key: county, unitary authority

But that's just the start of the jigsaw. For example England is divided into the **regions** on map **B**. These are in turn divided into smaller areas. Look at map **C**. Each area looks after its own services, such as education and social care.

About the UK

Some facts about the British Isles

	England	Scotland	Wales	Northern Ireland	Republic of Ireland
Flag of UK / Flag of Republic of Ireland					
Area (square kilometres)	130 400	77 100	20 800	14 200	70 300
Population (millions)	55.8	5.5	3.2	1.9	4.8
Flag of this British nation					

Remember!
The label matches the green area.

the British Isles

the United Kingdom

Great Britain (or just Britain)

History box (CE)

1801: Ireland becomes part of 'The United Kingdom of Great Britain and Ireland'.

1922: the Republic of Ireland gains independence. Northern Ireland remains in the UK.

1171: King Henry II of England invades and takes control of parts of Ireland.

1100: England, Scotland, Wales and Ireland are separate countries.

1276: King Edward I of England invades and takes control of Wales.

1536: King Henry VIII unites England and Wales, and makes himself King of Ireland.

1707: England, Scotland and Wales become 'Great Britain'.

Today: England, Scotland, Wales and Northern Ireland are still united as the UK.

Your turn

1. a How many countries are in the British Isles? Name them.
   b How many nations form the UK? Name them.

2. a Make a table with headings like this:

Region of England	A settlement in this region

   b In the first column list the regions of England. (Map **B**.)
   c In the second column try to name one town or city in that region. Use the map on page 139 to help you.

3. England's regions are divided into smaller parts. Map **C** shows the South West region. Its counties have existed for centuries. The unitary authorities are less than 40 years old.
   a Name three counties in the South West region.
   b Name two unitary authorities in the South West region.
   c Suggest a reason why the South West region has been divided up. (Is a big place easy to run?)

4. Table **D** uses simple blocks to show the British Isles.
   a Copy the table, including the blocks. (Draw them roughly!)
   b On your copy, colour in the blocks, or parts of them, to match the three titles. (Check the yellow panel above.)
   c Calculate the populations and areas to match the titles, using the data in the grid above. Add them to your table.

5. Why are there four nations in the UK? A timeline will help you find the answer. It's best to use a ruler to draw it.
   a Draw a line to represent the years 1100 – 2100 CE. (Would 10 cm work?) Put a mark every 100 years and label it.
   b Now draw arrows roughly at the dates given in the history box above, and add notes to show the events.
   c Add small maps or flags or other symbols to your timeline if you wish. Give it a suitable title.

6. London is the UK's capital. But Scotland, Wales and Northern Ireland also have capital cities. Unjumble their names, and identify the nation each belongs to! (Page 139?)
   a STAFBLE   b BINERDUGH   c DAFFRIC

D

	Great Britain	United Kingdom	British Isles
Population (millions)			
Area (sq km)			

3.3 What's our weather like?

Here you'll learn about weather patterns across the British Isles.

What is weather?

Weather means the state of the atmosphere. Is it warm? wet? windy?

Look at the weather map **A** on the right, for a day in October. Using the key, you can say that around place **P** that day:

- it was quite cloudy and wet, but there was some sunshine.
- the temperature was around 6 °C. Warm coats were needed!
- there was a south west wind (it blew *from* the south west).
- the wind was quite strong (around 38 miles per hour).

Our weather is changeable!

You know already that our weather can change from day to day. It can be different in different places on the same day, as weather map **A** shows.

Which parts are colder? warmer?

Although the weather can change from day to day, there are **patterns**. For example, some areas are *usually* colder than others. Look:

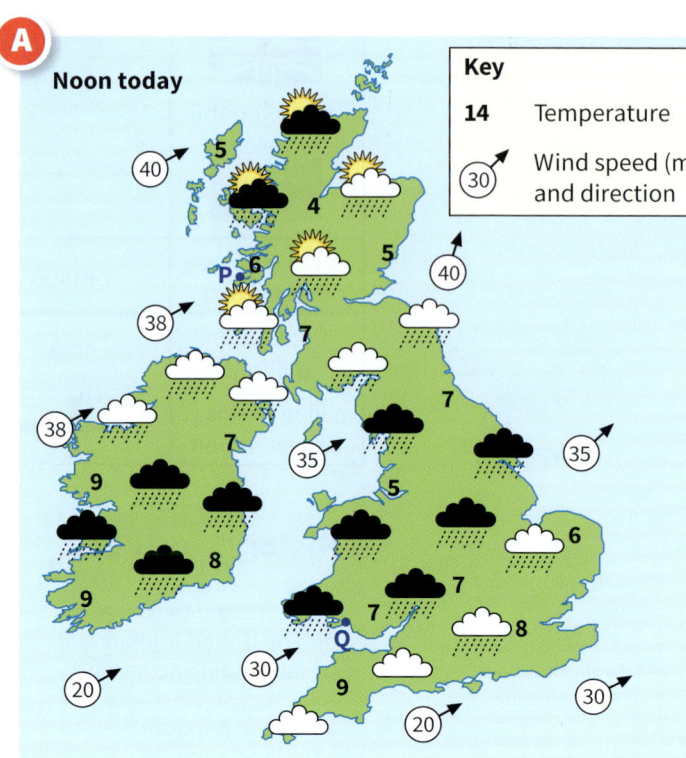

A Noon today

Key
14 Temperature
30 Wind speed (mph) and direction

B Temperature patterns in the British Isles

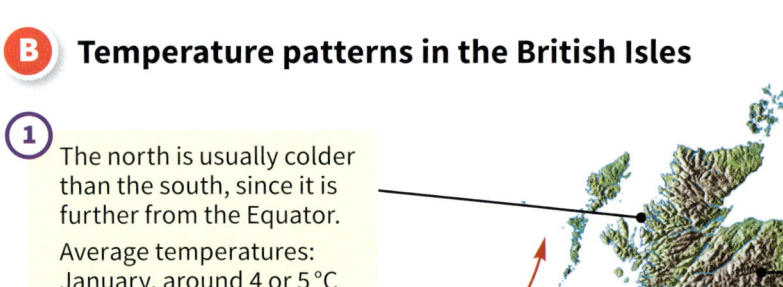

① The north is usually colder than the south, since it is further from the Equator.
Average temperatures:
January, around 4 or 5 °C
July, around 13 or 14 °C.

② It's colder on high land than on low land. The higher you go, the cooler it gets.

④ The west coast is warmer than the east coast, in winter. It is warmed by a warm ocean current called the **North Atlantic Drift**.

warm ocean current

③ The south is warmest since it is nearest the Equator.
Average temperatures:
January, around 6 or 7 °C
July, around 16 or 17 °C.

What if...
...it were warm and sunny every day?

Why...
...does it get colder as you go up a mountain?

46

About the UK

Which parts are wetter?

There is a pattern of rainfall too. Map **C** shows that some parts of the British Isles get a lot more rain than others.

Overall, the west of the islands and the upland areas are wetter. Follow the numbers below to see why:

C Average rainfall in one year

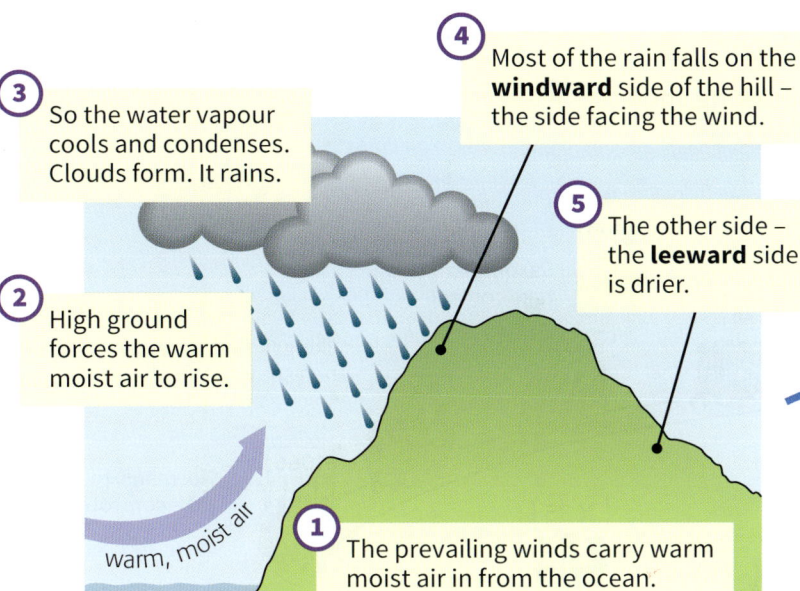

3. So the water vapour cools and condenses. Clouds form. It rains.

4. Most of the rain falls on the **windward** side of the hill – the side facing the wind.

5. The other side – the **leeward** side – is drier.

2. High ground forces the warm moist air to rise.

1. The prevailing winds carry warm moist air in from the ocean.

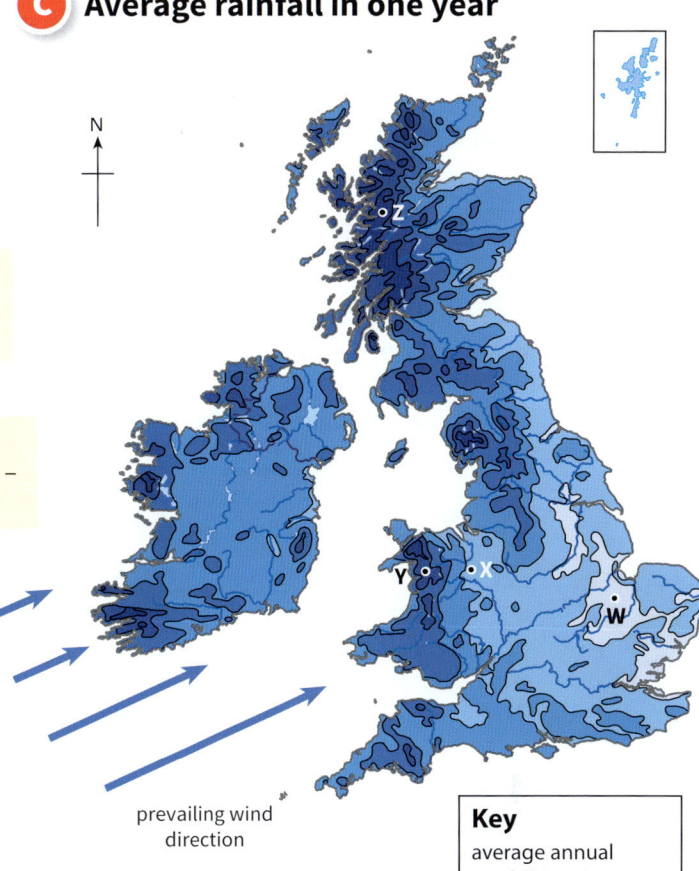

prevailing wind direction

Key
average annual rainfall (mm)

2400
1800
1200
800
600

Weather and climate

As you saw, **weather** means the state of the atmosphere at any given time. **Climate** is different. It is about the *pattern* of weather. It tells you what the weather is *usually* like in a place, at that time of year.

Overall the UK has an **equable** climate – it does not get too hot, or too cold. But now there is **global warming** – Earth is warming up. So the UK's climate is changing. We can expect hotter summers, wetter winters, and more storms.

Your turn

1. What's the weather like where you are today? Describe it. You might be able to use some of these words:

    sunny   cloudy   rainy   dry   calm
    cold   warm   mild   windy stormy

2. Look at weather map **A**.
    a Find the place marked **Q**. Say what the weather was like around **Q** that day, as fully as you can. Include wind!
    b Explain why it is usually warmer at **Q** than at **P**.

3. Look at map **C** above.
    Four places are marked on it: **W**, **X**, **Y** and **Z**.
    a Which one of the four is the wettest?
    b Which is driest?
    c Which one may have an average annual rainfall of:
        i 2000 mm?   ii 500 mm?

4. a What are *prevailing winds*? (Glossary.)
    b In the UK, the prevailing winds blow in from the south west. (Look at the arrows in map **C**.) They carry lots of moisture. Why? (Page 141?)

5. Explain the part that mountains play, in rainfall.

6. a Give *two* reasons why **Y** on map **C** gets a lot of rain.
    b **X** on map **C** gets far less rain than **Y**. Why?

7. a Explain the difference between *weather* and *climate*.
    b The UK has an *equable* climate. What does that mean?

8. In the UK, summers are now getting hotter and drier.
    a Why is this happening?
    b Suggest one way in which these changes could affect:
        i farmers in the UK   ii you

3.4 Who are we?

 Here you will learn that we're all descended from immigrants.

The long march

An **immigrant** is a person who moves here from another country, to live. 20 000 years ago, nobody lived here. (Much of the land was covered in ice.) So we are *all* descended from immigrants, if you go back far enough.

Over the centuries, many groups of people arrived. This drawing shows only the main groups. People are still arriving – and leaving too.

Did you know?
- London is less than 2000 years old.
- Damascus, the capital of Syria, is over 10 000 years old.

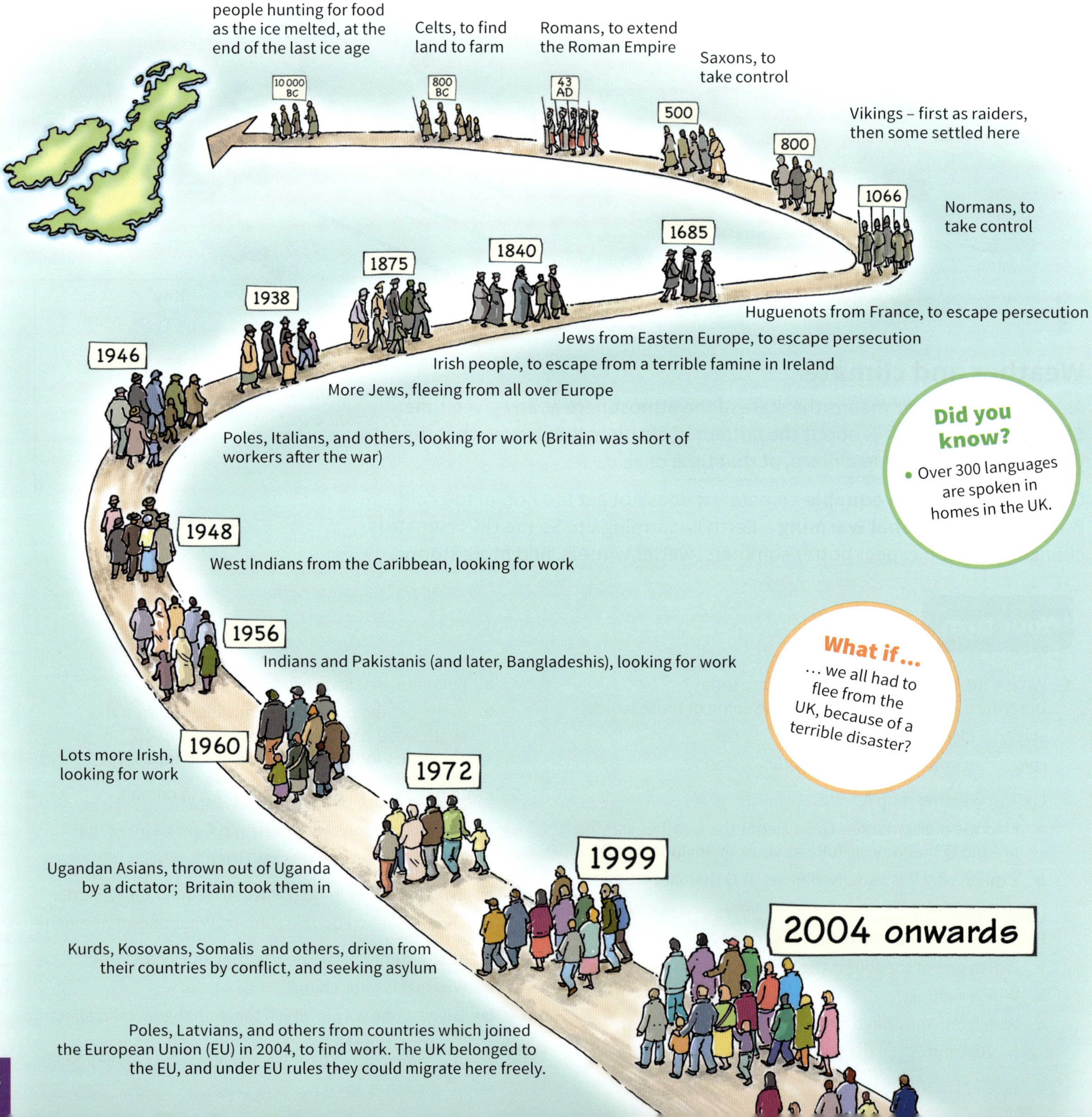

- 10 000 BC — people hunting for food as the ice melted, at the end of the last ice age
- 800 BC — Celts, to find land to farm
- 43 AD — Romans, to extend the Roman Empire
- 500 — Saxons, to take control
- 800 — Vikings – first as raiders, then some settled here
- 1066 — Normans, to take control
- 1685 — Huguenots from France, to escape persecution
- 1840 — Irish people, to escape from a terrible famine in Ireland
- 1875 — Jews from Eastern Europe, to escape persecution
- 1938 — More Jews, fleeing from all over Europe
- 1946 — Poles, Italians, and others, looking for work (Britain was short of workers after the war)
- 1948 — West Indians from the Caribbean, looking for work
- 1956 — Indians and Pakistanis (and later, Bangladeshis), looking for work
- 1960 — Lots more Irish, looking for work
- 1972 — Ugandan Asians, thrown out of Uganda by a dictator; Britain took them in
- 1999 — Kurds, Kosovans, Somalis and others, driven from their countries by conflict, and seeking asylum
- 2004 onwards — Poles, Latvians, and others from countries which joined the European Union (EU) in 2004, to find work. The UK belonged to the EU, and under EU rules they could migrate here freely.

Did you know?
- Over 300 languages are spoken in homes in the UK.

What if...
...we all had to flee from the UK, because of a terrible disaster?

All mixed up

We all carry the genes of past immigrants in our cells. How exciting! Look at these.

- Descended from Danu, a female Celtic warrior who was killed in a battle.
- Descended from Dayib, a Somalian trader who sold spices to the Romans.
- Descended from Olaf, a Viking wood carver who fell for a girl from Nottingham.
- Descended from Claudius, a Roman commander who lived in York.
- Descended from Gytha, a wealthy Saxon woman who owned 500 cows.
- Her grandparents came here sixty years ago, from a small village in India.
- Descended on his mum's side from Anne, a Huguenot silk weaver who fled to London.

What about you? Who are you descended from?

Your turn

1. What is an *immigrant*?

2. A **push factor** is anything that makes you want to leave a place. A **pull factor** attracts you to a place.
   Using the drawing on page 48 to help you, identify:
   a. four push factors that can drive people out of their countries
   b. three pull factors that have attracted people to the UK

3. Look at these four terms:
   A refugee   B invader
   C economic migrant   D asylum seeker
   a. First, define each term. (Glossary.)
   b. Then choose a group of immigrants from the drawing on page 48, to match each term.

4. When people move to a new country, they bring aspects of their culture with them. Give six examples of things from different cultures that you encounter in the UK. (Think about food, music, religious buildings, festivals, and so on.)

5. The graph on the right shows how the UK's population grew from 1930 to 2018. Some growth was due to immigration. More was due to **natural increase** – people having babies.
   a. Describe the shape of the graph.
   b. Overall, is the population rising or falling?
   c. What was the population (roughly) in:
      i. 2000?   ii. 2018?
   d. i. Around which year did the population begin to fall?
      ii. Give a reason for this fall in population. (History!)

6. People leave the UK too. They emigrate.
   a. Define the term *emigrate*. (Glossary.)
   b. Look at these numbers for the UK for 2017:
      immigrants 601 000   emigrants 331 000
      Calculate the increase in the UK population that year, through migration.

7. Imagine you are 18. And you are going to emigrate from the UK. That's brave!
   a. Which country will you choose? List the things (pull factors) which attract you to that country.
   b. Predict the difficulties you may face:
      i. in the UK before you leave
      ii. in your new country
      You could write two lists, or draw a spider map. Don't forget your feelings!

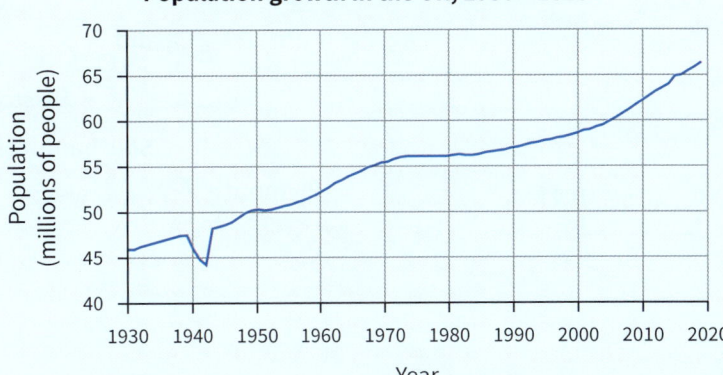

Population growth in the UK, 1930 – 2018

3.5 Where do we live?

 Here you'll see how we are spread unevenly around the UK.

Population density

The **population** of a place means the number of people who live in it. The population of the UK is around 67 million.

Population density is the average number of people living in an area, per square kilometre. Map **A** shows how this changes across the UK.

Look at the key. The darker the shade, the more people per square kilometre. The palest areas are the **least populous**.

B The UK's 10 largest cities

	Name	Population (millions)
1	London	8.84
2	Birmingham	1.14
3	Leeds	0.78
4	Glasgow	0.62
5	Sheffield	0.58
6	Manchester	0.55
7	Bradford	0.54
8	Edinburgh	0.50
9	Liverpool	0.49
10	Bristol	0.46

A [Map of the UK showing population density with cities labelled: Inverness, Aberdeen, Y, Glasgow, Edinburgh, Belfast, Newcastle, Bradford, Manchester, Leeds, Liverpool, Sheffield, Coventry, Birmingham, Cardiff, Bristol, Oxford, Cambridge, Z, London, Dover, Southampton, Portsmouth, Plymouth, X]

Key
Population density (people per square km)
- more than 200
- 101–200
- 51–100
- 5–50
- fewer than 5
- country capital
- city
- border between countries
- border between nations

Did you know?
- The UK is the 21st biggest country in the world, by population.

Did you know?
- The UK has 274 people per sq km, on average.
- France has 119.
- The USA has 35!

▲ Rural Wales. The village is Brechfa.   ▲ An urban area in England: Birmingham.

About the UK

E Where the UK population lives

83% urban areas
17% in rural areas

F Great Britain's coalfields

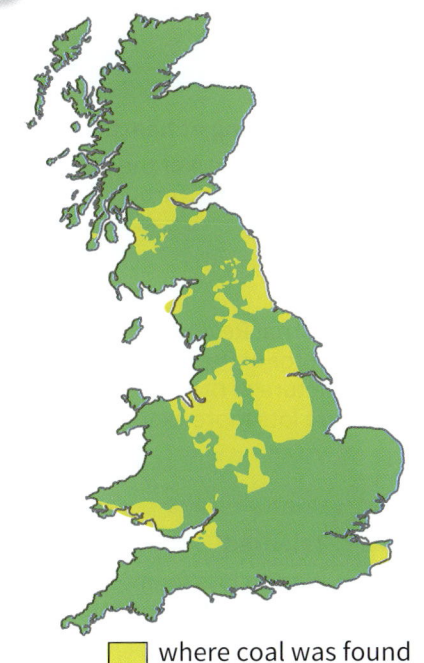

☐ where coal was found

Urban or rural?

Photo **C** shows a **rural area**. A rural area is mainly countryside, but it may have villages and small towns. Photo **D** shows an **urban area**. Urban areas are built up. They include larger towns, and cities.

Now look at pie chart **E**. 83% of the UK's population lives in urban areas.

Why are people spread unevenly?

So why are people spread unevenly in the UK? Mainly because of physical geography! It is not easy to build a town, or earn a living, up in the mountains.

During the Industrial Revolution, settlements grew fast in areas with easy access to coalfields, and to rivers for water and transport. (Coal was burned as fuel in factories, and used in extracting iron, as you saw on page 12. The iron was used for railways, ships, and other things.)

The coast has always been important too, for fishing, sea transport, and trade.

Your turn

1. What does *population density* mean?

2. a Describe the population density in the place on map **A** that is labelled : i X   ii Y   iii Z
   Use the correct term *people per square km* in your answer.

   b Explain why the population density around **Y** is low. The map on page 139 will help.

3. Using table **B**, calculate how many more people there are in London than in: **a** Birmingham **b** Edinburgh

4. For the cities shown on map **A**, identify:
   a two cities in northern Scotland
   b one city in Wales
   c five cities on the coast
   d the city almost directly south of Cambridge
   e the city south west of Bristol
   f the city less than 50 km from Birmingham
   g the city nearly 200 km north east of Manchester

5. **C** shows a rural area. **D** shows an urban area. **E** shows that most people in the UK choose to live in urban areas. Why? Suggest *at least* five reasons. (Think of things people need, and like to do.) Draw a spider map?

6. Compare Great Britain in maps **A** and **F**.
   a Can you find any correlation (link) between coalfields and population density? Describe what you notice.
   b Explain any correlation that you find.

7. Use what you've learned in this unit to write a report called *The pattern of population density in the UK*.
   - Make it at least 10 sentences long. (Try for more!)
   - Use as many terms as you can from the white box below.
   - The map on page 139 will help.

mountainous	capital city	England	most populous
urban areas	coast	least populous	Scotland
Wales	Northern Ireland	coalfields	rural areas

3.6 How are we doing?

Here you'll look at some different aspects of the UK.

If the UK were a person …
Imagine the UK is a person. What is that person like?

1 Quite old – over 40!
Of the 67 million people in the UK, just over half are aged over 40.

2 But young at heart …
Almost 1 in 4 of the population is under 20 years old. (Around 16 million people.)

3 Smart!
The Industrial Revolution began here. The inventor of the internet was British. The structure of DNA was worked out here. There are lots more examples.

4 Talented!
World-class for music … fashion … the media … computer games … and more.

5 Sporty!
World-class in some sports. (Could do better in others.)

6 Loves …
jeans and trainers, a cup of tea, takeaways, ice cream vans, animals …

7 Loved by …
tourists from other countries. Around 40 million a year come to visit!

8 Works hard
Works in factories and on farms, producing things to sell. But mostly offers services – like teaching you, looking after you when you're ill, serving you in cafés …

9 Sells things to other countries
Sells oil, cars, chemicals, aircraft, medical drugs …
And services such as banking and insurance, entertainment, tourism.

10 Buys things from other countries
For example oil, coal, gas, cars, steel, goods like iPads and computers, and lots of food and clothing.
And services too …

11 Doing fine, thanks!
Out of 193 countries, the UK usually ranks in the top 25, for average wealth per person.

12 But still a little anxious!
Loses sleep over some things. Like …
- how to improve healthcare
- crime
- terrorism

You'd look so much better with fur.

About the UK

Not the same everywhere

Overall, the UK is doing fine, compared with most countries. But it's not the same story all over the UK. There are big differences. Some areas are wealthy, with people earning lots. Others are run down, and people may not be able to find any work. (For example, local factories may have closed.)

What if… …everyone in the UK earned the same amount?

Your turn

1. Look at page 52. See if you can pick out five facts about the UK that you did not know before. Write them down, *in your own words!*

2. See how many examples you can give, of British musicians who are stars around the world. (Singers and/or groups.)

3. Box **5** is about sport. Which sports does the UK excel in, in your opinion? Which sports could we improve in?

4. Box **7** is about tourism. What a lot of tourists! What do you think attracts them to the UK? You could show your answer as a spider map.

5. Box **8** is about the work we do. Around 30 million people go out to work, in the UK. You know some of them! List at least 15 jobs that people do.

6. Jobs can be put into groups or **sectors**. Box **C** shows the three main sectors.
   a. Think about each job in your list for question **5**. Then write **P**, **S** or **T** beside it, to show the sector you think it belongs to. (**P** for primary!)
   b. Look at your answers for **a**. Did you have jobs for each sector? Which sector had most?

7. a. Name three types of goods the UK *exports*.
   b. Now name three types of goods the UK *imports*.

8. Compare the places in photos **A** and **B**.
   a. Which place looks wealthy? Give your evidence.
   b. The other place is quite run down. How can you tell?
   c. When a factory closes, an area may become run down. Explain why.

9. The *average* pay in 2018 for people in the UK was **£539 per week**. Now look at the average pay within these cities:

Cardiff	£505 per week
Edinburgh	£598 per week
London	£727 per week
Manchester	£512 per week

   a. In which of the four cities was the pay:
      i above the UK average?   ii below the UK average?
   b. Calculate the difference in average pay per week between Manchester and London.
   c. Suggest some reasons why the average pay in London is so high. (Page 54 may help!)

C Job sectors

Primary: people take things from the earth and the ocean. Examples: *farmer, fisherman, miner*.

Secondary: people make things in factories, or construct things on building sites. Example: *builder*.

Tertiary: people provide services for other people. Examples: *doctor, teacher, taxi driver, hairdresser*.

3.7 London, our capital city

Here you'll learn about London, and how its population has grown.

London, a global city

Photo **A** shows London, the UK's capital city, at night. From about 400 km up. Look at the River Thames, snaking across the city on its way to the sea.

- London is the UK's biggest city by far, with around 8.8 million people. Amost one in every eight people in the UK lives in London.
- It's a **multiracial** and **diverse** city, with over 300 languages spoken.
- The British government is based here.
- It's known around the world for fashion, music, theatre, art, and shopping. It has world-famous football teams too. (Can you name any?)
- It is a **global city**, which means it has impact far beyond the UK. In particular it is a world hub for finance (banking, insurance, the stock market and so on).

▲ *London at night, photographed from the International Space Station. The brightest bridge across the river is Tower Bridge. Look for a red dot further west on the river. That's the London Eye.*

How did London begin?

It began with the Romans! In 47 CE, four years after they invaded Britain, the Romans built a bridge across the River Thames. They brought goods up the river by boat, and unloaded them by the bridge. Soon a small settlement sprang up around the bridge. They called it **Londinium**.

Londinium grew fast, through trade with mainland Europe. So the Romans made it their capital.

When the Roman army left Britain, Londinium went downhill. But by 600 CE it had started to grow again. It has been growing for most of the time since.

▲ *Great public transport.*

▲ Lots to look at.

▲ Lots to do.

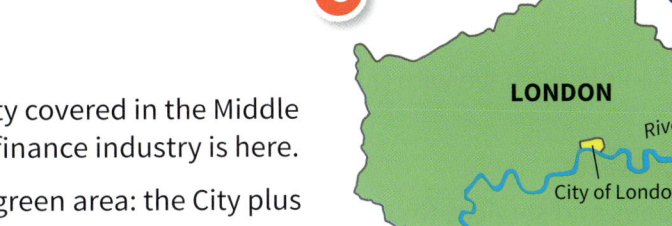
B

C

London today

Look at map **C**. The **City of London** is the area the city covered in the Middle Ages. It is still called the City, and much of London's finance industry is here.

But London grew and spread, and now it covers the green area: the City plus 32 **boroughs**. Each borough is run by its own council.

The full name for the green area is **Greater London**, although we usually just call it **London**. Map **B** shows the counties that border it.

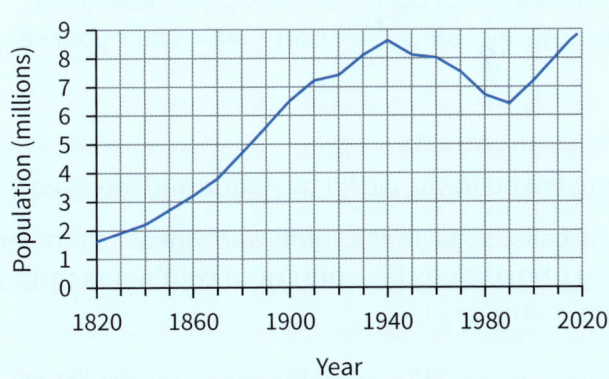

Your turn

1 How many famous buildings or structures in London can you name? Write a list.

2 Where is London? Describe its location in the UK in five sentences. (Check the map on page 50, and map **B** above.)

3 About how wide is London (Greater London) at its maximum?
   a 9 km   b 29 km   c 59 km
   Use the scale in map **C** to help you choose.

4 Now look at photo **A**.
   a What is that dark band wiggling through the photo?
   b What do all the orange lines show?
   c What do the pale cream splodges show?
   d The photo has many other dark shapes, besides the river. Unjumble these terms to discover what some of them are:
      i *karps*   ii *veresoirrs*   iii *sprots tadiumss*

5 London is a major transport hub, with road and rail links to other parts of the UK, and air links to other countries. What evidence is there in photo **A** that London is a transport hub?

6 From photo **A**, identify what you think is the busiest area of London. Then describe where it is in London, and give at least two reasons to justify your choice.

7 This graph shows London's population from 1820 to 2018.

   a *London's population has increased non-stop since 1820.* True or false? Give evidence for your answer.
   b i Around which year did the population start to decline?
     ii Suggest a reason for the decline. (History!)
     iii When did the population start to grow again?

8 Today, people move to London from all over the UK, and from other countries. Suggest three **pull factors** that attract them.

9 Thanks mainly to its location, London has grown far larger than any other British city. Which aspects of its location have helped it to grow? (The maps on pages 139 – 141 may help.)

3.8 Our links to the wider world

 The UK has many links to the rest of the world.

Our links with the rest of the world

Look at all the links we have with the rest of the world.

1 Trade
- We sell goods and services to countries all over the world, and this earns us £billions.
- We also buy goods and services from other countries.

2 Transport
- We can fly abroad through 25 British airports.
- The Channel Tunnel links us by rail to France.
- The UK has lots of ports. Around 30 link us to other countries – including ferry ports, which take your car.

3 Communications
- We are linked to the world by phone and internet.
- Phone calls, texts, music, films and videos: all travel to and from islands (like ours) and continents through cables on the sea floor.

4 Investment
- Lots of British companies have been bought by companies from the USA, China, India, France, and other countries.
- Foreign companies now supply much of our water, gas, and electricity, and run railways, buses, and ports.
- In the same way, British companies buy companies in other countries.

5 Membership
- The UK belongs to several 'clubs' of countries. For example the **Commonwealth** and the **United Nations**.
- We have also signed many international **treaties** – for example about human rights, and the environment.

6 Tourism
- Around 40 million tourists from other countries visit the UK each year. They spend billions of pounds.
- At the same time over 70 million tourists from the UK visit other countries each year.

7 Culture
- British music, fashion, theatre, films and books make an impact around the world. So do our football teams.
- Some of our TV programmes are sold all over the world.
- It works the other way too. We absorb culture from other countries, and especially the USA.

8 Aid
- Every year, the UK gives 0.7% of its **gross national income** – the total money it earns – as **aid** to poorer countries in Africa and Asia.
- That's 7p out of every £100.
- Some is for dealing with disasters – like earthquakes. Some is for education, to give young people a chance.

▲ It's trade! The world's largest container ship (400 m long) at the port of Felixstowe in Suffolk.

About the Commonwealth and the United Nations

The Commonwealth
- The Commonwealth has 53 member countries. Most were once British colonies. India, Pakistan, Nigeria, and Australia are members.
- The Commonwealth countries are linked through their shared history and values, and use of English as an official language.
- Quite a lot of trade goes on between them.

The United Nations (UN)
- This body was set up at the end of World War II, to promote world peace.
- It also tackles other issues. For example it works to help refugees.
- Nearly all countries belong to it.
- The UK is on its **Security Council**. This decides whether to intervene in conflicts around the world. It can impose **sanctions** on a country. (A sanction is a penalty – for example a ban on trade with that country.)

Links with the European Union
- The **European Union** or **EU** is a union of European countries which trade freely together, without **tariffs** (taxes). People can also move freely between the countries, to work.
- By 2016, the EU had 28 member countries, including the UK.
- But in June 2016, the UK voted to leave the EU, by a small majority. The government struggled to agree a good **Brexit** (British exit) deal. We missed our leaving deadline in March 2019.
- Did we finally leave the EU? When? Discuss with your teacher!

▲ *The Commonwealth Games are held every four years: Australia in 2018, the UK in 2022.*

▲ *The UN headquarters is in New York.*

Your turn

1. The UK has many different kinds of links to the rest of the world. Unjumble these four!
   a *tuulrec* b *snattorrp* c *dreta* d *stumior*

2. Trade with other countries is very important to the UK. Suggest a way in which trade with other countries benefits:
   a factories in the UK
   b shoppers in the UK, like you

3. You want to visit Paris. Describe two ways you could travel there, using different forms of transport.

4. Do you ever link to other countries through undersea cables? If yes, give one example.

5. Two of the links on page 56 are very important for allowing trade and tourism to take place. Identify them, and justify your choice.

6. Why would a foreign company want to buy a water or electricity company in the UK? Suggest a reason.

7. The UK belongs to these two organisations:
   a the Commonwealth
   b the United Nations
   Describe each one, in your own words.

8. The UK is on the Security Council of the United Nations.
   a Explain why this is an important role for the UK.
   b The Security Council can impose sanctions on a country which is misbehaving. What are *sanctions*?

9. What % of the money it earns each year does the UK give in aid to poorer countries?

10. Think about this opinion. Do you agree with it? Decide, and write at least five sentences to justify your decision.

It would be best if the UK cut its links with the rest of the world.

3 About the UK

Let's see how much you've learned about the UK ...

check ✓

1. The pale yellow areas on map **A** represent the British Isles.
   a. Name the country or nation labelled with this number:
      i 1   ii 3   iii 5   iv 2   v 4
   b. Name the entity that is made up of these areas:
      i 2 + 3 + 4 + 5   ii 3 + 4 + 5
   c. The area labelled 6 is part of another country. Which country?
   d. The dots labelled **v, w, x, y** and **z** represent the capital cities of the different countries and nations.
      i Identify each city. Write your answers like this:
         w = _____
      ii Underline the names of the two cities which are country capitals.
   e. The average population density in area 3 is about 67 people per sq km. In area 4 it is it about 395 people per sq km. Give one reason to explain this difference.

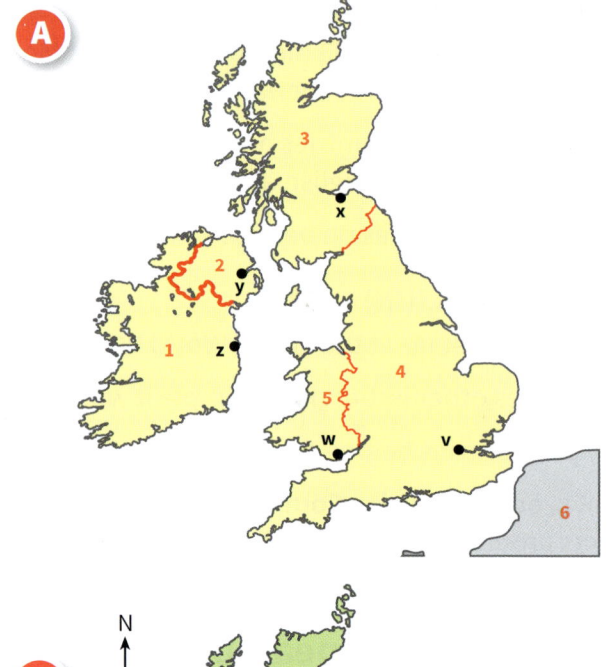

2. This table is about the climate for the three places marked on map **B**. These places are also on the map on page 139.

Place	Average temp (°C)	Annual rainfall (mm)	Altitude (m)
Swansea	10.4	1361	14
Southend-on-Sea	9.9	572	29
Dalwhinnie	6.6	1304	351

   a. Define these terms: i prevailing wind   ii altitude
   b. i Which of the three places gets most rain?
      ii Give one reason why this place has a wet climate.
   c. i Calculate how much more rain falls on Swansea than on Southend-on-Sea, in a year.
      ii Suggest a reason why Southend-on-Sea has quite a low annual rainfall.
   d. Southend-on-Sea is a seaside resort. Suggest:
      i one positive impact   ii one negative impact
      of the low rainfall on the people who live there.
   e. Give *two* reasons why the average temperature is lower in Dalwhinnie than in Swansea.

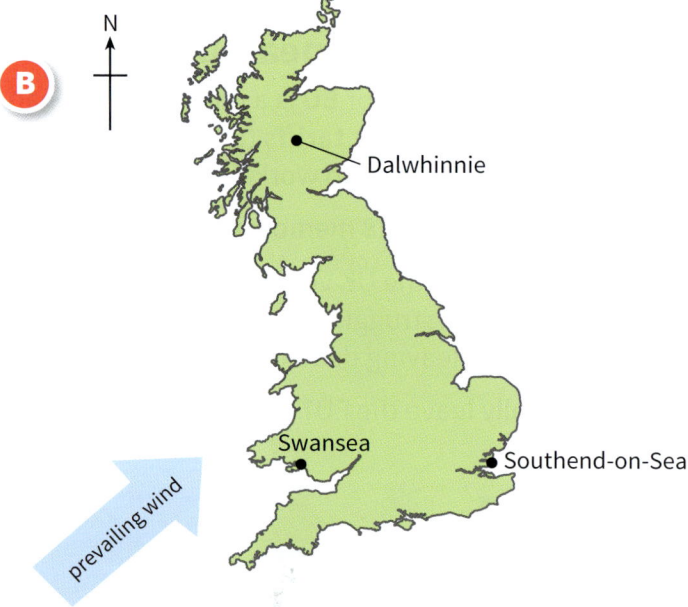

3. Pie chart **C** shows what the workers in the UK do for a living. The three sectors are explained on page 53.
   a. Give two examples of jobs in each sector.
   b. Decide whether this statement is true, or false:
      i There are more people working on farms than in factories, in the UK.
      ii Most workers in the UK provide services.
   c. Name three types of goods that the UK exports.
   d. Suggest two benefits for the UK if we made more goods for export.

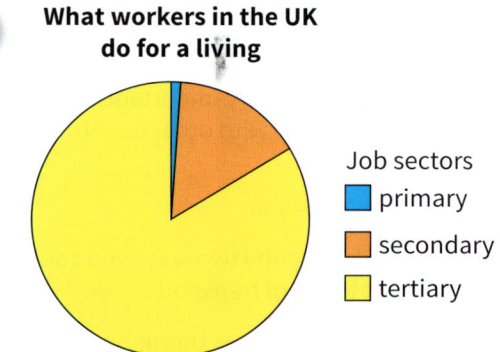

4. *Our links with other countries help to make the UK a better place to live.*
   Do you agree with this statement? Decide, then write at least half a page to justify your decision.

4 Glaciers

4.1 Your place … 20 000 years ago!

 Find out what your place was like, and why, 20 000 years ago.

It's time to travel

Climb into your time machine. Press the button. And whoosh! Travel back in time, to see your place … as it was 20 000 years ago.

What will you find? It depends on where in the UK you live!

If you live in a white area on map **A** below, you'll find a thick sheet of ice when you arrive. There are no humans or other animals. No grass. No trees. It's brutally cold.

If you live in the grey area, there is no ice sheet. But it's very cold, and there's snow. There are no humans – but you may see woolly mammoths, and bison!

Why was it like that?

Why was your place like that, 20 000 years ago?

Because around 110 000 years ago, Earth got colder and colder. A new **ice age** began. (There had been many others before it.)

Over time, an ice sheet spread over much of northern Europe, and most of the British Isles. Look at map **A**.

The ice sheet did not reach the grey areas in map **A**. But these areas were still very cold. The ground was deeply frozen. In summer the surface thawed, giving thin boggy soil on which small plants grew. This type of environment is called **tundra**.

By 10 000 years ago, Earth had warmed up again. The ice age ended. The ice over the British Isles melted away. And today we have ice for only short times, in winter.

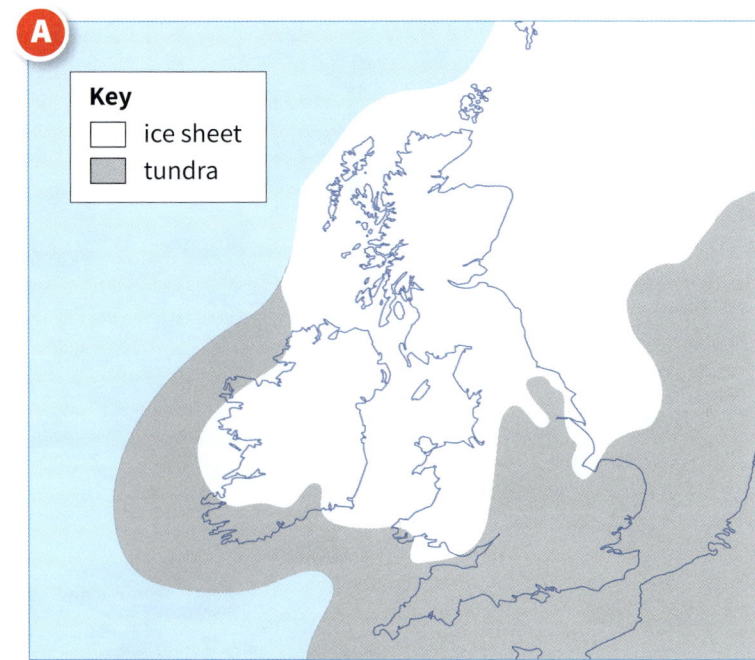

Glaciers

There was more land then!

During the ice age, water levels in the ocean were much lower than today. (Up to 120 m lower.) That's because so much water was locked up in ice. The water drained away from shallow parts of the ocean floor, and they became land.

Look at map **B**. It shows that when the water levels fell, the British Isles were joined by land to the rest of Europe!

What about people?

20 000 years ago, there was nobody in the British Isles.

We had arrived here earlier in the ice age – 40 000 years ago. We had just walked across from other parts of Europe. But as the ice sheet spread, it got too cold for us. We left.

Then about 12 000 years ago, when the ice sheet was shrinking, we came back to the British Isles again.

The animals

But there *were* animals here, 20 000 years ago. There were woolly mammoths and bison and Arctic foxes, which could survive the tundra winter.

And in summer, when plants grew in the tundra, large herds of reindeer and antelope walked over from other parts of Europe, to feed.

When the ice age ended

As Earth warmed up again, the ice melted. So water levels in the ocean rose again. And about 8 100 years ago, the rising water cut us off from the rest of Europe.

But the ice had changed the landscape – and we can still see the results today. You'll find out more in later units.

▲ Many fossils of mammoth tusks have been found in the UK and the North Sea.

Your turn

1 Define these terms: **a** ice age **b** tundra (Glossary?)
2 About how long did the last ice age last? When did it end?
3 **a** The British Isles were joined by land to the rest of Europe, in the last ice age. (Map **B**.) Explain how this happened.
  b Later they separated from the rest of Europe again. Why?
  c Suppose they had not separated. How would your life be different today?
4 Get ready! You'll travel back 20 000 years, in your home place – and stay for three days. You can take only 20 items.
  a First, pick out where you live, on map **A**. Is it in the white part or the grey part? (The map on page 139 may help.)
  b Now write a list of what you will take with you. Beside each item, explain why you will take it.
  c Good. You've arrived! What's it like there? Write a blog for us! At least five sentences.

4.2 Glaciers: what and where?

 Here you'll find out where the ice is on Earth today – and start learning about glaciers!

What are glaciers?

During the last ice age, ice covered about a third of Earth's land. Today it covers about a tenth.

The ice does not just sit there. It flows! We call it **glaciers**.

Glaciers are large masses of ice that flow across the land, and down slopes. Giant glaciers, that cover huge areas, are called **ice sheets**.

▲ *Ice over Earth during the last ice age.*

Where are the glaciers?

As you'd expect, glaciers are found in Earth's coldest places. There are some on every continent – even Africa! Look at this map:

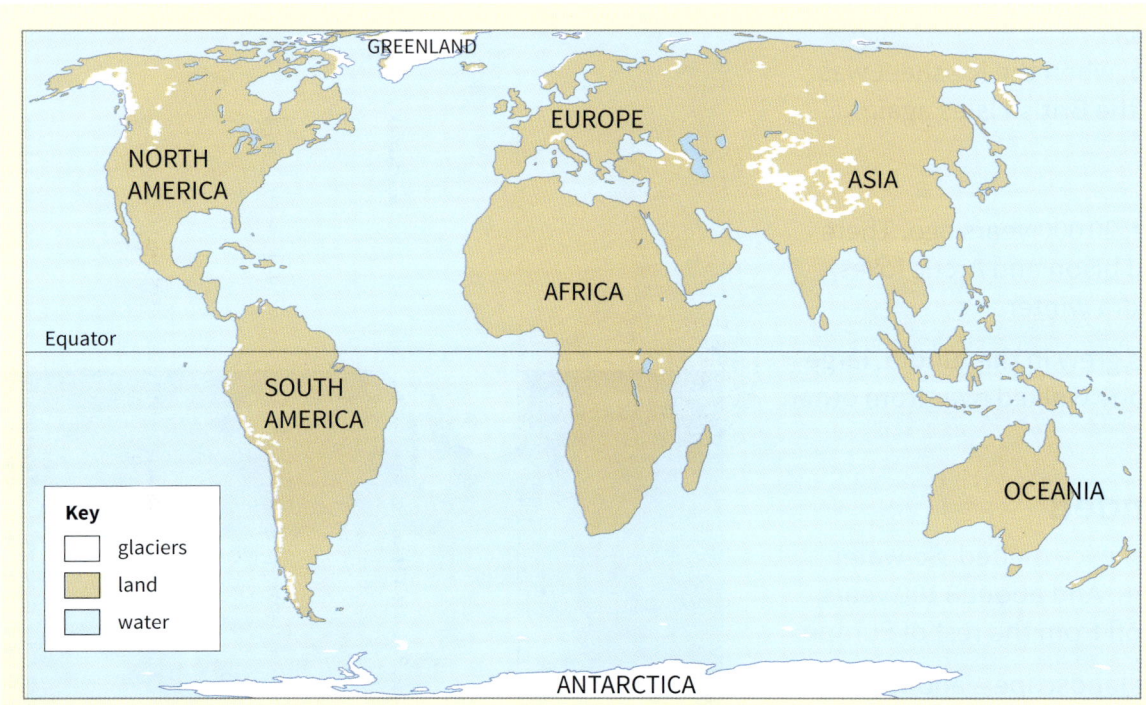

Did you know?
- 75% of the world's fresh water is frozen in glaciers.

Why…
…is it called Greenland?

What if…
…all the glaciers melted?

Far from the Equator, at the top and bottom of the world, ice sheets cover Antarctica and most of Greenland. Between them, they have over 99% of Earth's ice. They are more than 4 km thick in places. Picture that!

Earth's other glaciers are much smaller. Most are high up in mountains, where it's also very cold. Most big mountain ranges have glaciers. They are called **mountain glaciers** or **alpine glaciers**.

Glaciers depend on snow

Rivers depend on **rain** falling from the sky. Glaciers depend on **snow**!

In those cold places, snow falls layer upon layer. Over time, the layers below get compacted to ice, like when you squeeze a snowball very hard. It could take a layer of snow 10 metres thick to make a layer of ice 1 metre thick.

As it gets thicker, the ice gets heavier and heavier. And eventually it starts to flow. A glacier is born!

Did you know?
- Glaciers are only on land.
- The ice that forms when an ocean freezes over is called sea ice.

Glaciers

Glaciers flow

Glaciers don't just sit there. They flow. How?

First, ice can flow *inside* a glacier, because the ice crystals slide over each other, under pressure. And second, the ice at the bottom of the glacier can melt due to the weight of the glacier; then the whole glacier slides along on the water.

Ice sheets flow just a few metres a year. Mountain glaciers flow faster down their slopes – 300 metres a year or more.

Where do they flow to?

A mountain glacier flows down the side of the mountain, in a valley. And eventually it reaches a place where it melts.

In ice sheets, the ice flows out to the thinnest parts, like when you pour syrup. In Antarctica, it flows into the ocean in places, and floats as an **ice shelf**. Bits of the ice shelf break off now and then to form **icebergs**.

▲ Watching that river of ice flow by – so slowly. The Aletsch Glacier in Switzerland.

▲ Looking down on Antarctica from space. The flat parts with blue dots marked on are ice shelves, where the ice sheet flows into the water.

▲ Made it! Cracks or **crevasses** form where the ice gets stretched. For example where the glacier flows round a bend.

Your turn

1. Explain how a glacier forms. You could draw diagrams – beginning with a snowfall – and add notes to them.
2. What is the difference between a glacier and an ice sheet?
3. Outline the two ways in which glaciers flow.
4. a The UK has no mountain glaciers today. Why not?
   b Identify five countries which do have mountain glaciers today, using the maps on pages 62 and 140 – 141.
5. You are a scientist. Your job is to study the glacier in photo **B**.
   a How could you prove that it is flowing? Suggest a method.
   b How would you calculate how fast it is flowing?
   c The middle of a glacier flows faster than the edges. Suggest a way to prove this too, for your glacier.
6. Crevasses can open up quickly – and close up quickly too. You are in photo **C**. You fall into the crevasse. It is 20 m deep. It creaks loudly. It is closing! What happens next?

63

4.3 How do glaciers shape the land?

 In this unit you'll find out how glaciers shape the land they flow over.

Glaciers work as they flow

As you saw on page 63, glaciers don't just sit there. They flow. And as they flow, they scrape and shape the landscape, like giant bulldozers.

They do three jobs:
1. They pick up or **erode** material.
2. They carry it away, or **transport** it.
3. Then they drop or **deposit** it.

Let's look at these processes in more detail, for mountain glaciers.

Scrape and shape, 24/7.

1 Erosion

Glaciers pick up material in two ways.

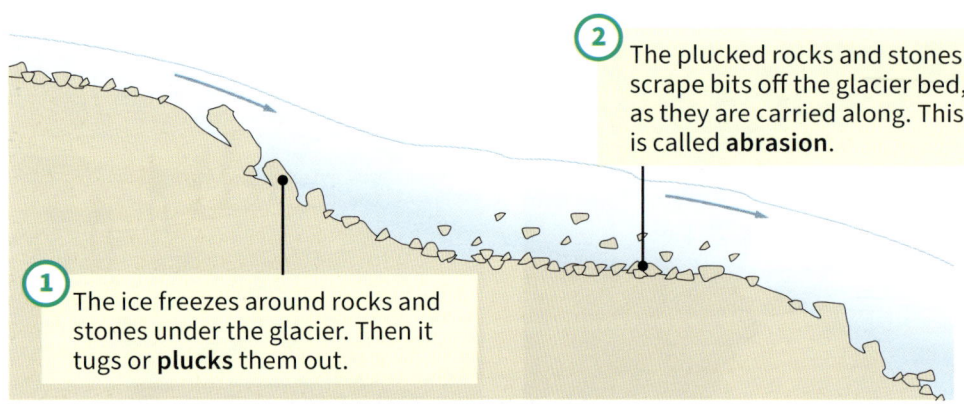

① The ice freezes around rocks and stones under the glacier. Then it tugs or **plucks** them out.

② The plucked rocks and stones scrape bits off the glacier bed, as they are carried along. This is called **abrasion**.

A process called **freeze-thaw weathering** makes plucking easier by breaking up rock. First, water freezes in cracks in the rock. As it freezes it expands, making the cracks wider. Later, the ice thaws. The cracks fill with water again. It freezes again. The cracks get even bigger. And so on … until the rock breaks into sharp pieces.

2 Transport

The glacier then carries away the material it has eroded. This drawing shows a slice through the glacier. Look how the material is carried.

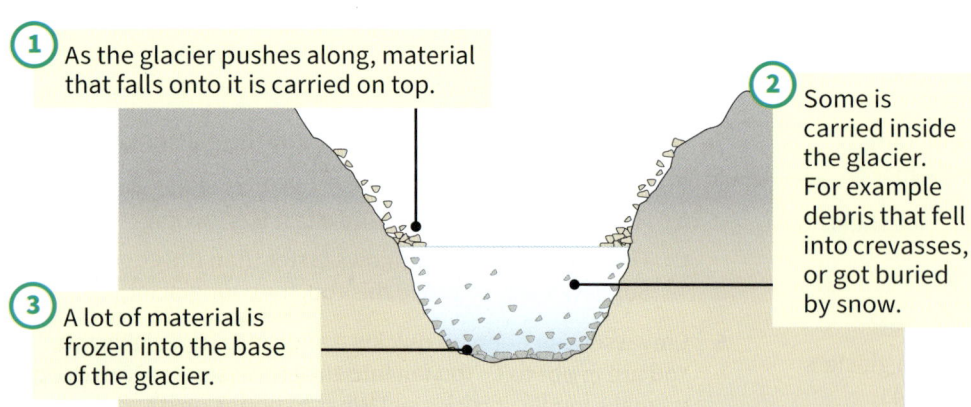

① As the glacier pushes along, material that falls onto it is carried on top.

② Some is carried inside the glacier. For example debris that fell into crevasses, or got buried by snow.

③ A lot of material is frozen into the base of the glacier.

▲ *A glacier went this way, once upon a time! These deep scratches in the rock are the result of abrasion. They are called **striations**.*

▲ *Another glacier in the Alps. Look at the debris on it. Glaciers often look very dirty.*

3 Deposition

As you go down a mountain, it gets warmer. So eventually the front of the glacier reaches a place where it melts. Look at **C**.

As the ice melts, the load it is carrying falls to the ground – rocks, stones, sand, and clay, all mixed up together. This mixture is called **till**.

The water from the melting ice is called **meltwater**. It runs off, and will feed a river or lake.

Meanwhile, higher up the mountain, snow keeps on feeding the glacier. So the glacier keeps on flowing down to the place where it melts.

▲ *An Alpine glacier melting. The end of a glacier is called its **snout**. The meltwater looks milky, because it carries lots of tiny particles produced by abrasion.*

Glacial landforms

The result of all this work by glaciers is **glacial landforms**. (Landforms are features of the landscape.)

There are glacial landforms throughout the UK, in the areas that were **glaciated** during the last ice age. (Look at the white areas in map **A** on page 60.) The result is some spectacular scenery – and especially in high or **upland** areas.

There are many good examples in the **Lake District** in England. So we will refer to the Lake District often in the rest of this chapter.

Words to remember

glacial – to do with glaciers
That's a glacial landform!

glaciated – covered and shaped by glaciers, now or in the past
Most of Ireland was glaciated during the last ice age.

glaciation – the process or results of being covered by glaciers
We're studying glaciation this week.

Your turn

1 A glacier is like a great big bulldozer. Explain why.
2 Name the erosion process through which:
   a glaciers pick up chunks of rock as they move along
   b glaciers scrape the ground they pass over
3 Describe the rock in photo **A**, and explain how it got those deep scratches. Name the scratches.
4 Freeze-thaw weathering breaks rock into pieces. It occurs where water repeatedly freezes in rock – even where there are no glaciers. Drawing **D** shows the process.
   a Make a larger copy of the drawing, and add its text. (You do not need to copy the rocks exactly!)
   b Complete the sentences for steps 2 and 3, to explain what's going on. Page 64 will help.
   c Now explain why this process makes plucking easier, when a glacier is flowing along.
5 Look at all the debris on top of the glacier, in photo **B**. How did it get there?
6 Now look at photo **C**.
   a Where is the liquid coming from?
   b Name the liquid, and explain why it looks milky.
7 It's time to start your own glossary about glaciation. To make your glossary:
   a list all the words you met about glaciation so far
   b beside each word, write its definition.

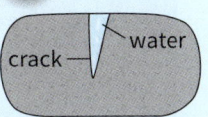 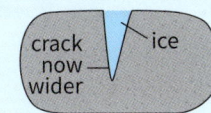 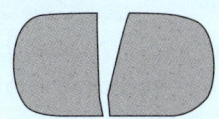

1 Water seeps into a crack in the rock.
2 As the water freezes it expands, so …
3 Steps 1 and 2 are repeated, and eventually …

4.4 Landforms shaped by erosion – part 1

Here you'll get an overview of the glacial landforms shaped by erosion – and a closer look at three of them.

First, an overview

Let's compare a landscape before and after it had glaciers, to see how they changed it.

1

Look at this landscape. A mountain, rivers, and valleys carved out by the rivers. This area has a mild climate. It gets plenty of rain – but very little snow.

2

Then the climate changes. Heavy snow falls year after year. Lower down, most of it melts away again. High in the mountain, it builds up. Mountain glaciers start to form.

3

Thousands of years later, the ice age has settled in. The glaciers have grown, and flow down the valleys. They are scraping and shaping the land as they go.

4

Now the ice age is over. The glaciers have melted. But they have left a changed landscape, with jagged features. All the features named here were created by **erosion**.

A

▲ A corrie in the Lake District. It is shaped like an armchair. The tarn in it is called Blea Water.

▶ Another corrie, with Bleaberry Tarn. The long lake is Crummock Water.

B

Glaciers

More about corries, arêtes and pyramidal peaks

The last drawing on page 66 showed the three features below. Check it again!

Corrie

A **corrie** begins as a sheltered hollow, where snow builds up year after year.

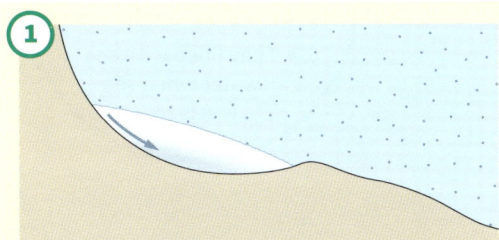

The snow compacts to ice. When the ice is thick enough, it starts to flow. First it flows within the hollow.

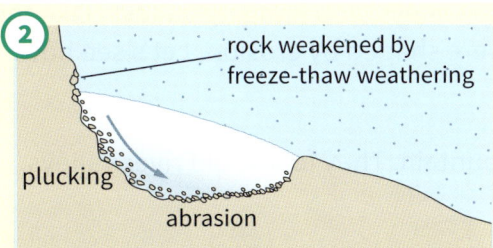

Plucking and abrasion make the hollow deeper, and the walls steeper. Freeze-thaw weathering helps.

Now the glacier is big enough to flow over the edge of the corrie. It heads down the mountain.

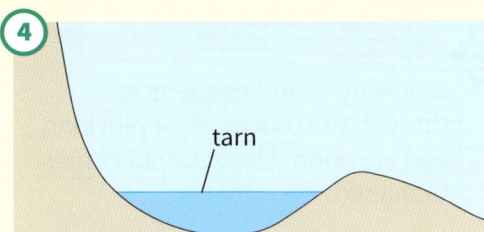

When the glacier melts, the corrie is revealed. It may have a lake in it. Corrie lakes are often called **tarns**.

▲ This famous arête in the Lake District is called Striding Edge. The lake on the left is called Red Tarn.

Arête

Sometimes two corries form side by side. The glaciers erode the rock between them, leaving a sharp ridge of rock. It is called an **arête**.

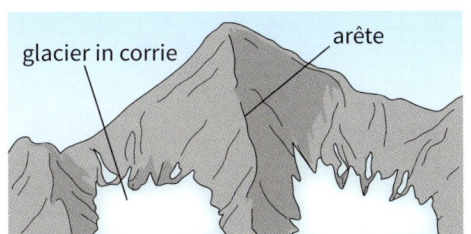

Pyramidal peak

Imagine three or four corries around a mountain top. Their glaciers cut back into the mountain top, leaving a **pyramidal peak**.

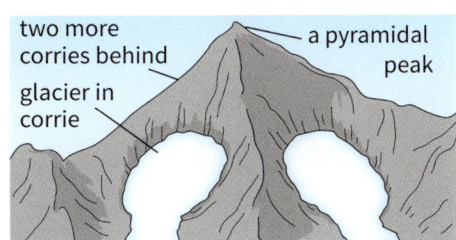

▲ A famous pyramidal peak: the Matterhorn. It's in the Alps, on the border between Italy and Switzerland. It has four faces. Over 500 climbers have died on it.

Your turn

1 Drawings **1** and **4** on page 66 show a landscape before and after glaciation. Compare them, and identify at least three ways in which glaciation has changed the landscape. You could give your answer as bullet points.

2 Now select photo **A**.
  a Draw a sketch of the scene, and label the features.
  b Draw a set of diagrams to show how the corrie formed.

3 Next, choose photo **C** or **D**. Draw a sketch of it, and add labels and clear notes to explain how the landform (arête or pyramidal peak) was formed.

4 Imagine you are standing on top of the landform (arête or pyramidal peak) in **C** or **D**. Look around. What do you see? How do you feel? Write your answer as a blog for a travel website. Include the name of the landform, and the place!

4.5 Landforms shaped by erosion – part 2

 Let's look at two more glacial landforms from drawing 4 on page 66.

Two more landforms shaped by erosion

Drawing **4** on page **66** showed U-shaped valleys, and hanging valleys. Like the other landforms in the drawing, they were shaped by erosion. Let's see how.

U-shaped valley

Glaciers take the easy route down a mountain. They follow old river valleys.

> **Did you know?**
> - The fjords of Norway are U-shaped valleys carved out by glaciers.
> - They filled with sea water when sea levels rose.

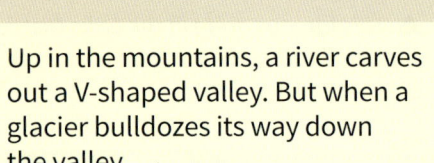

| Up in the mountains, a river carves out a V-shaped valley. But when a glacier bulldozes its way down the valley … | … it widens and deepens it, through the processes of plucking and abrasion. The valley becomes U-shaped. | When the glacier melts, a river may flow again. Now it's in a wide valley that it did not create. It is called a **misfit** river. |

Compare the two photos below. Photo **A** shows a V-shaped river valley. Photo **B** shows a U-shaped valley carved out by a glacier.

A U-shaped valley is a clear sign that a glacier has passed through. The Lake District has lots of U-shaped valleys. (They are also called **glacial troughs**.)

▲ A V-shaped valley in Wales, cut out by the River Twymyn.

▶ The U-shaped valley seen from Newlands Hause in the Lake District. The little misfit river to the right of the road is Keskadale Beck.

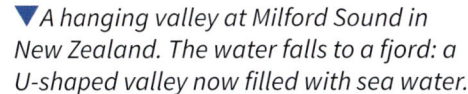

Glaciers

◀ Two ribbon lakes in the Lake District: Buttermere (in front) and Crummock Water.

▼ A hanging valley at Milford Sound in New Zealand. The water falls to a fjord: a U-shaped valley now filled with sea water.

Now look at the photo above. It shows two *lakes* in a U-shaped valley. Long thin lakes like these are called **ribbon lakes**.

Imagine a glacier scraping along the valley. It reaches a place with softer rock, so it digs this out more deeply, making a trench. When the glacier melts, the trough fills up with water. That's how a ribbon lake forms.

Hanging valley

A **hanging valley** is a small valley that hangs above a larger one.

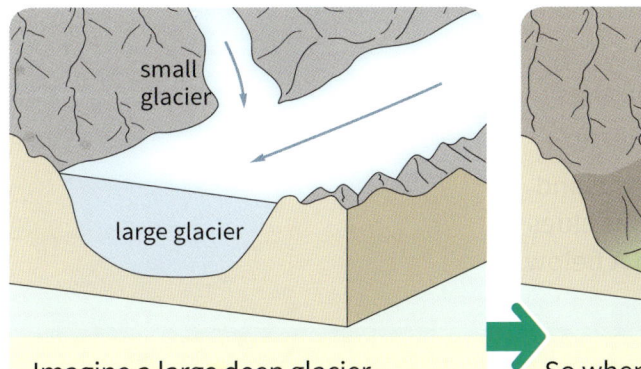

Imagine a large deep glacier moving along a valley. A smaller one joins it. It is much less deep.

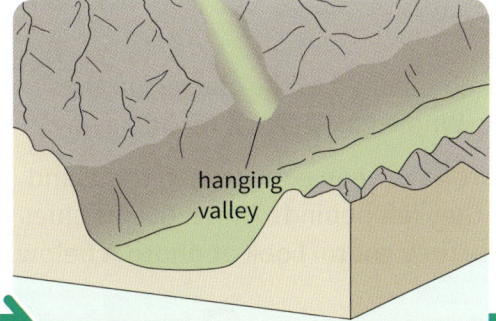

So when the ice melts, it reveals the smaller valley hanging above the larger one.

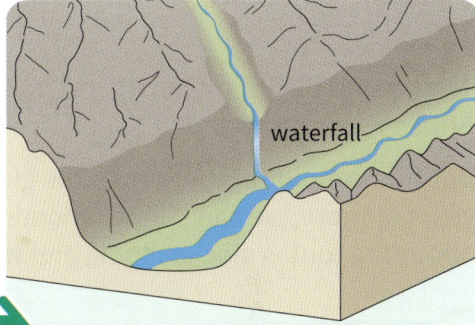

If a river flows in the smaller valley, it will splash into the larger valley as a waterfall. (Look at photo **D**.)

Your turn

1 a Identify the landform in photo **B**.
   b How was this landform formed? Explain in 25 – 30 words. (Your own words, not copied from the page.)
   c Now draw a sketch from the photo, and add notes and labels. Don't forget the river and road.

2 *U-shaped valleys show the power of glaciers in shaping the land.* Write a paragraph to justify this statement. It might help to imagine the glacier that formed the valley in photo **B**.

3 a The valley in photo **D** is called a *hanging valley*. Why?
   b Outline how a hanging valley is formed. Briefly!

4 a The lakes in **C** are called *ribbon lakes*. Why?
   b The drawings in **E** show how a ribbon lake forms. Make larger copies. Add labels and notes to explain what's going on.

5 Choose photo **B**, **C** or **D**. Write an ad for a travel website to attract tourists to that place.

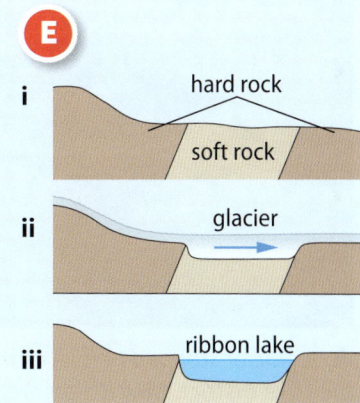

69

4.6 Landforms created by deposition

  Here you will learn about landforms created when a glacier melts.

Moraines

As you go down a mountain, it gets warmer. So as a glacier flows down a mountain, it reaches a point where it will melt. But it may melt even at the top of the mountain if the climate warms up! Look at these diagrams.

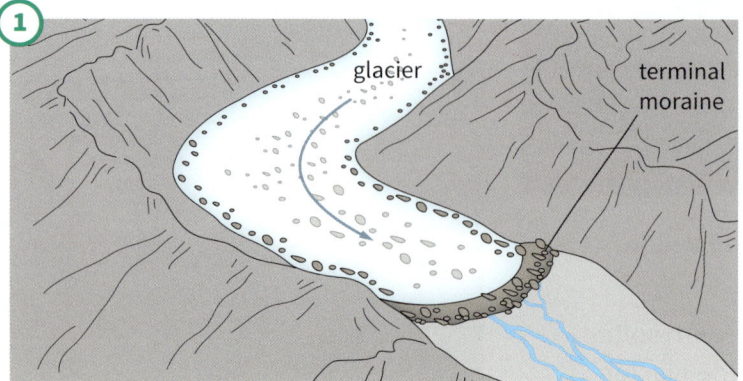

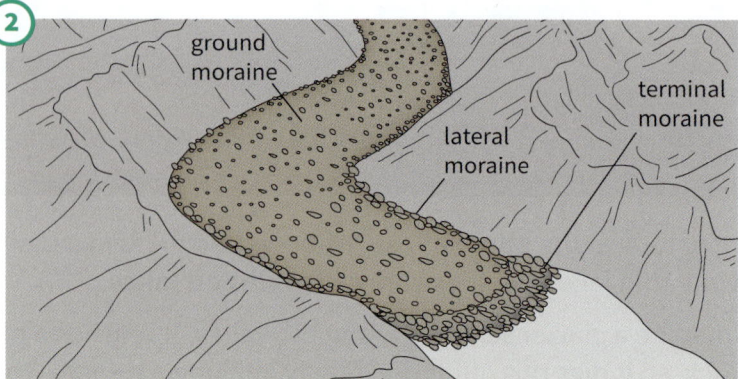

A glacier flows non-stop, carrying its load of rocks, stones, sand, and clay. This mixture is called **till**. When the front reaches a place where it melts, the till is deposited. It builds to form a ridge called a **terminal moraine**. (*Terminal* means *at the end*.)

But suppose the whole glacier melts. The till that's on top, along the edges, also drops to the ground. It forms a ridge called a **lateral moraine**. (*Lateral* means *side*.) The till that is frozen into the base gets spread all over the valley floor, as **ground moraine**.

The second diagram above sums up what happened to the glaciers in the British Isles, at the end of the last ice age. Whole glaciers just melted away.

But think about this. Although the glaciers have been gone for ten thousand years or more, we can still see moraines they left behind! These give us clues about the routes glaciers took, and where they got to. Look at photo **A** below.

▲ A glacier stopped here! A terminal moraine at Borrowdale in the Lake District. Now it's covered in grass and ferns and bushes.

▲ Ground moraine is a thick layer of till deposited along the melted glacier's route. Today it may be gently rolling farmland.

Glaciers

Erratics

A glacier can carry huge rocks. When it melts, the rocks are dropped.

They may be a long way from where they started – and very different from the other rocks around them. They look clearly out of place.

These stray rocks are called **erratics**.

I got dumped.

▶ *An erratic in a field in Kentmere, in the Lake District.*

Drumlins

Drumlins are another sign that an area has been glaciated.

Drumlins are low hills, shaped like the back of a spoon.

Experts are not sure how they formed. But all agree that the smooth shape is due to a glacier flowing over till that had already been deposited.

▶ *Drumlins in the Yorkshire Dales – not far from the Lake District.*

Your turn

1 Define these terms: a till   b moraine
   Answer in full sentences. The glossary may help.

2 Name the type of moraine that is found:
   a only along the sides of a glacier's route
   b all along a glacier's route

3 Look at the bank in photo **A**.
   a Explain how this bank formed.
   b What would you expect to find, if you dug down into it?
   c Why might the farmer on the right feel this way about it?

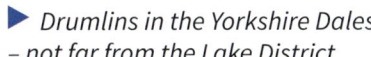
I wish that glacier had stopped somewhere else!

4 Now look at photo **C**.
   a How did this big boulder get here?
   b An *error* is a mistake. Boulders like this one are called *erratics*. See if you can explain why.

5 Drumlins are shaped like the back of a spoon. Look at this diagram.

glacier flowed this way

   Now look at photo **D** above. In which direction did that glacier travel, thousands of years ago?
   a from X to Y   b from Y to X

6 Moraines, erratics and drumlins have something in common. They are the result of *tpoionesdi* by *csgleari*. Unjumble the two jumbled words.

7 And now … if you began your own glossary for glaciation (in question 7 on page 65), it's time to update it.

71

4.7 More about the Lake District

Here you'll learn more about the Lake District, an upland area in north west England. And you'll explore an OS map.

The Lake District

The Lake District is in Cumbria, north west England. It is a **National Park**. That means it is protected for us all to enjoy – thanks mainly to its stunning scenery.

And this is in turn thanks to glaciers! During the last ice age, glaciers pushed their way down its slopes. The Lake District photos on pages 66 - 71 show some of the landscapes they sculpted.

The glaciers are gone – but not forgotten. Today, 10 000 years later, they still influence how people in the Lake District earn a living!

Glaciation on the OS map

The OS map opposite shows part of the Lake District. Most contour lines are very close together. This means the slopes are steep. The numbers show they're high.

The contour lines will help you pick out the landforms the glaciers left behind. Look at these two examples.

▲ A conifer plantation at Thirlmere in the Lake District. (Conifers bear cones.)

U-shaped valley

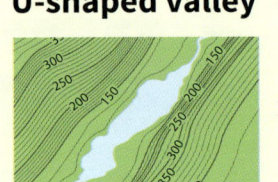

1. The sides of the valley are steep, so the contour lines are close together.
2. But the bottom of the valley is quite flat, so the contour lines are far apart.
3. There may be a ribbon lake in the valley – as here – or a misfit river.

Corrie

1. A corrie is rounded, so the contour lines are curved, a bit like a horseshoe.
2. Its sides are steep, so the contour lines are close together.
3. It may have a lake in it – which may be labelled 'tarn' on the map.

How people in the Lake District earn a living

The Lake District has a **population** of around 41 000. These are the three main ways the workers earn a living.

Tourism Most work in tourism. Tourists flock here to enjoy the scenery and outdoor activities. They need food, a place to sleep, and other amenities. Over 19 million tourists visited in 2017.

Farming It's mainly sheep farming. The soil was stripped off the slopes by the glaciers, and it is still thin today, with poor vegetation. The climate is also cool, with quite a lot of rain. But sheep can cope!

Glaciers deposited till on the low land, so the soil there is more fertile. Cattle are reared on the lowland grass, and some crops are grown.

Forestry Conifers cope well with the conditions on the slopes, so there are large plantations of these trees. They are used for timber and paper making.

▲ Young tourists about to take a boat trip on Lake Windermere in the Lake District.

Glaciers

Your turn

1. Photo **B** on page 68 shows a U-shaped valley with a beck.
   a. Find this valley on the OS map. (Look for flatter land!)
   b. Identify the square where the photographer stood to take the photo, and give its four-figure grid reference.
   c. In which direction did the photographer face?
      i south   ii north   iii north east
   d. What is a *beck*? The map and photo will help you answer.
   e. Keskadale Beck is a *misfit*. How can you tell from the map?

2. The map shows two lakes.
   a. They are r_____ lakes. Complete the word. (Page 69?)
   b. Which one is deeper? Give your evidence.
   c. About how long is Crummock Water, in km? (Scale!)

3. Now look back at photo **C** on page 69. The photographer stood in one of these squares to take it. Which one?
   a. 1616   b. 1914   c. 1815

4. a. Find a tarn on the map, and give its name.
   b. Name the type of landform the tarn sits in.

5. The term *force* is used for waterfalls, in the Lake District. Name and give four-figure grid references for two waterfalls.

6. Compare squares 1716 and 1417. In which square are you *less* likely to find sheep? Justify your answer.

7. Give the grid reference for a square with a conifer plantation. (Check the OS symbol for conifers on page 138.)

8. Pick out four different types of facility on the map, that help to attract or cater for tourists. (A camp site, for example? Page 138.) Give 6-figure grid references for them.

9. Describe the area shown on the map. Is it hilly? Crowded? What about rivers, forests, villages? How do people earn a living? You can use bullet points, or a spider map, to answer.

4.8 Do glaciers matter?

 Do glaciers matter? You can think about it here!

Do they affect us?
Earth still has glaciers, as you saw on page 62. But do they affect us? And do they matter? Read on …

1 Bringing in tourists

Tourists love glaciers. Like this one – the Perito Moreno in Argentina. And even where glaciers have long gone (as in the UK), tourists love the landscapes they left. Tourism helps places because it brings in money and creates jobs.

2 Presenting a challenge

Many people dream of climbing Mount Everest, in the Himalayas. It is Earth's highest mountain. It sits on the border between Nepal and China. (And it's a pyramidal peak.)

The photo shows the Khumbu Icefall – like a waterfall, but it's a glacier. It is the most dangerous part of the climb. If you want to climb Everest, or other high mountains, prepare for glaciers. Bring your ice axe!

3 Supporting life

People depend on glaciers for survival in some places – and especially in Pakistan.

The great River Indus runs down through Pakistan. Up to half its water is meltwater from glaciers in the Himalayas. Farmers use the river water for their crops.

Several other major rivers are partly fed by glaciers. For example the Ganges in India, and the Yangtze in China. Millions of people depend on these rivers for water.

Glaciers

4 In need of protection

Antarctica has almost 90% of Earth's ice. The land below the ice sheet may be rich in minerals.

Nobody owns Antarctica. But seven countries claim slices of it: the UK, France, Norway, New Zealand, Australia, Chile, and Argentina. The USA and Russia have research stations there.

One day, these claims may cause conflict. But for now, mining is banned by the **Antarctic Treaty**. It protects Antarctica as a place of peace and scientific research.

5 Melting!

Glaciers are made of ice. So if Earth warms up, they melt. And today, Earth is getting warmer.

Scientists are watching the ice sheets closely. They appear to be melting already: they are getting thinner.

The trouble is, as they melt, the water level in the ocean rises. If all the Greenland ice sheet melts, the rise will be 7 metres. Coastal places will flood, affecting millions.

Mountain glaciers are shrinking too. Look at the photo on the left.

Earth has warmed up often before, for natural reasons. This time, most scientists agree that humans are the main cause. Mainly because of the fossil fuel (coal, oil, and gas) that we burn.

Your turn

1 *Glaciers help some people to earn a living.*
Give as many examples as you can, of people who depend on glaciers for a living. (Not only the farmer in box **3**!)

2 *Glaciers help people to enjoy life.*
See how many examples you can give this time.

3 Mount Everest is in the Himalayas, in Asia. If the Himalayan glaciers melt away, how might it affect:
   a children in Pakistan?   b climbers tackling Everest?

4 Seven countries claim slices of Antarctica (including the UK). Why would anyone want to own land that lies under an ice sheet? Suggest as many reasons as you can.

5 Look at the photo with the sign, above.
   a This glacier would have looked different in the year 2000. In what way?
   b The change is explained below, with some words missing. Copy and complete.

   It is getting _____ in the region where this glacier is. So _____ snow falls. That means there is _____ snow to feed the _____. So the glacier has _____.

6 See if you can explain these two statements.
   a People all over the world pose a threat to the ice sheets.
   b The ice sheets pose a threat to people all over the world.

4 Glaciers

How much have you learned about glaciers? Let's see.

check ✓

A

B

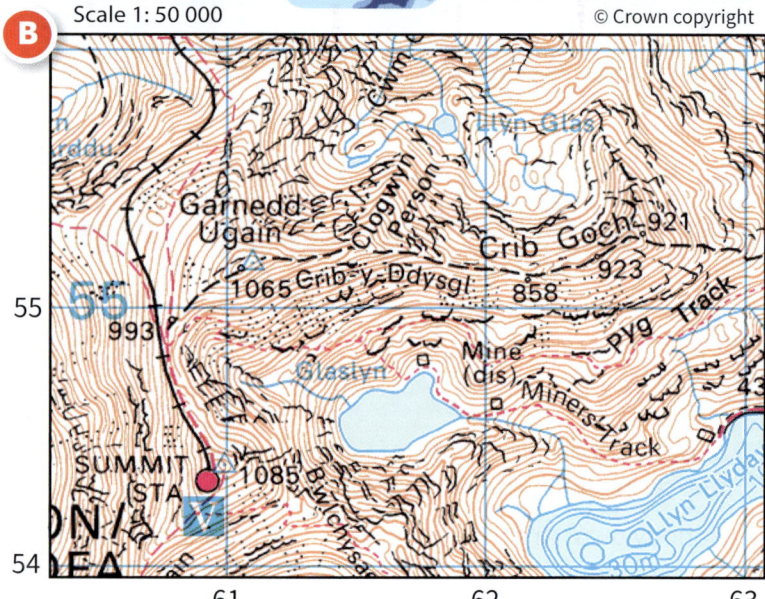

Scale 1:50 000   © Crown copyright

1 Photo **A** was taken in Snowdonia, a National Park in Wales. During the last ice age, glaciers shaped this landscape.
   a Identify the landforms with these labels, in photo **A**:
      i   W   ii   X   iii   Y   iv   Z
   b Name the two processes through which glaciers erode the landscape.
   c The hikers are climbing on rock that's broken up. Name a natural process which breaks up rock.
   d State two ways in which the landscape in **A** would have looked different, before glaciation.

2 OS map extract **B** includes the area shown in photo **A**.
   a Z in the photo is the peak of Mount Snowdon, the highest mountain in Wales. Give a four-figure grid reference for Snowdon. (Look for the largest spot height.)
   b The feature at Y in the photo is called Crib Goch.
      i   Give a four-figure grid reference for it.
      ii How can you tell from the map that Y has steep slopes?
   c The lake (tarn) in the photo is also on the OS map. Name it. (Look at the N arrow and rotate the photo in your mind!)

3 Snowdonia gets lots of visitors. Bar graph **D** shows visitors per month at the Visitor Centre, for a typical year.
   a In which month(s) did the centre have:
      i   fewer than 5000 vistors?
      ii more than 15 000 visitors?
   b i   Compare the visitor numbers for January and August. (Check page 16 for *compare*.)
      ii Suggest *three* reasons for the difference you found in **i**.
   c Photo **C** shows Snowdonia's mountain railway. Vistors buy tickets. Suggest one benefit of this railway for:
      i   visitors   ii   the managers of the national park
   d Using the map in **B**, give a 6-figure grid reference for the railway station near Mount Snowdon's peak.
   e Suggest one negative impact of this railway.

C

D

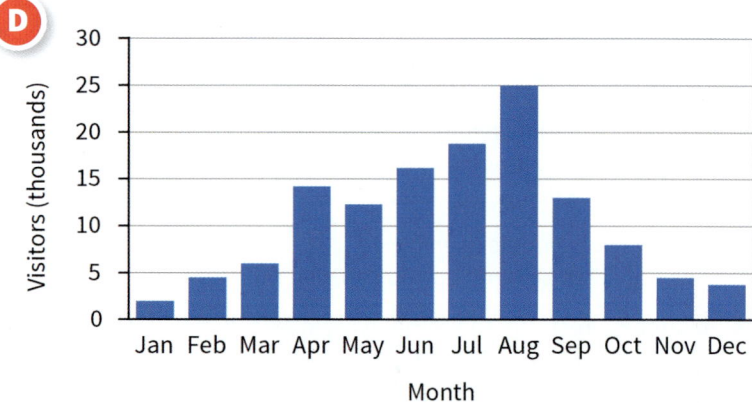

4 *There is a strong link between glaciation and tourism. Write half a page to explain this link, and include at least two examples from this chapter.*

5 Rivers

5.1 Meet the River Thames

Here you'll learn about England's longest river, and its journey from source to sea.

Did you know?
- The River Severn is the longest river in the UK.

Why... ...do we give rivers names?

It starts as a puddle!

Look at these photos of the River Thames.

This is the start or **source** of the river: a spring seeping up in a field in the Cotswolds. (Look at the map on the next page.) The site is called Thames Head.

Here is the Thames 20 km later, near Cricklade. A small shallow river, flowing through meadows. But it is on a mission. It is heading for the North Sea, over 300 km away.

This is Henley, half way on the river's journey to the sea. Between the source and here, fifteen smaller rivers have joined the Thames. No wonder it has grown!

This is the river 20 km further on, at Lechlade, on the edge of the Cotswolds. It has got much wider and deeper. And now it is deep enough for boats and barges.

And this is London! The Thames flows through the city. That's the Millennium bridge. The white dome belongs to St Paul's Cathedral. (See the green dot on the map.)

The journey's end. Over 50 km from the centre of London, and 346 km from its source, the Thames flows through the wide **Thames Estuary** and into the North Sea.

Rivers

A map of the river

This map shows the route of the River Thames, and some of the villages, towns, and cities that grew up along it. The smaller rivers that join it are called **tributaries**.

Thames Tideway

The photo on the right shows the Thames in London. Compare it with the matching photo on page 78.

Both show the same scene. But the water level is different. That's because the Thames is **tidal**, from Teddington – marked by a yellow dot on the map above – to the sea.

As the tide rises in the North Sea, sea water moves up the Thames Estuary towards Teddington. So the water level in the river rises. As the tide falls, water drains away again.

The tidal stretch of the river is called **Thames Tideway**. The water level here changes non-stop, from high to low tide and back. There are two high tides and two low tides each day.

▲ Low tide in London. Look at the pillars of the bridge. Seven hours from now, the water level will have risen by 6 or 7 metres.

Your turn

1. Give six facts about the River Thames. Include facts about its length, and location in the UK.
2. Now draw a sketch map for the River Thames. Do not show its tributaries. But do mark in and label:
   - the Cotswolds, the range of hills where it rises
   - at least six settlements (cities, towns, villages) along it
   - the Thames Estuary, and the North Sea
   - the tidal stretch of the river. (Use a different colour?)
3. See if you can explain these facts.
   a. The Thames has much more water in it by the time it reaches London, than it had at Lechlade.
   b. The water in the Thames in London is a bit salty.
   c. The Thames pours over 60 million cubic metres of water into the North Sea each day – and still does not run dry.
4. Look again at the map above. The Thames wiggles a lot on its journey to the North Sea. Suggest a reason!

5.2 It's the water cycle at work

Without the water cycle there'd be no rivers – and no you. Here you'll find out why.

Did you know?
- The rain that falls on you has fallen millions of times before.
- It may have fallen on a dinosaur.

What is the water cycle?

Water sloshing around in the sea this week may rain down on you next week. It's the **water cycle** at work. Follow the numbers …

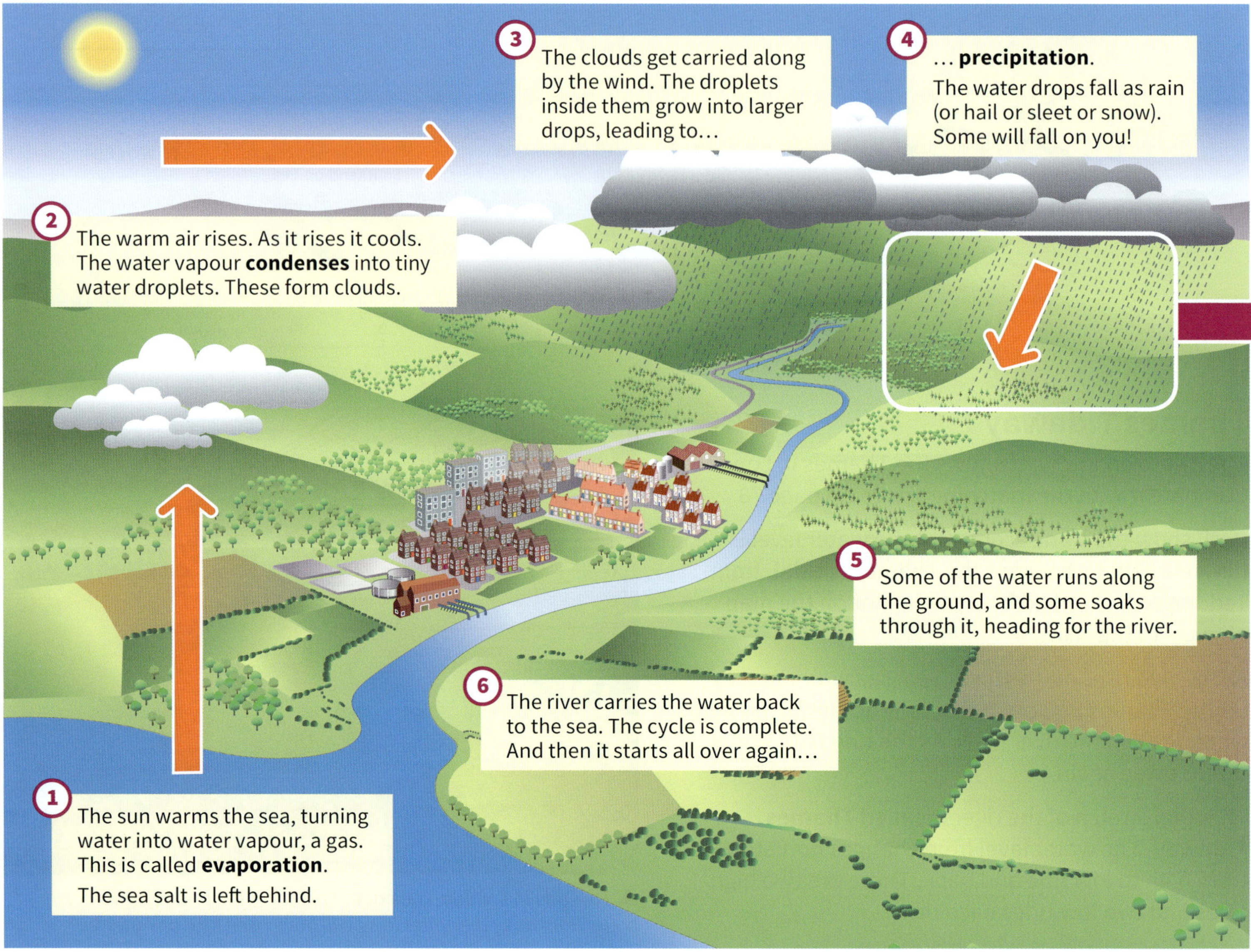

① The sun warms the sea, turning water into water vapour, a gas. This is called **evaporation**. The sea salt is left behind.

② The warm air rises. As it rises it cools. The water vapour **condenses** into tiny water droplets. These form clouds.

③ The clouds get carried along by the wind. The droplets inside them grow into larger drops, leading to…

④ … **precipitation**. The water drops fall as rain (or hail or sleet or snow). Some will fall on you!

⑤ Some of the water runs along the ground, and some soaks through it, heading for the river.

⑥ The river carries the water back to the sea. The cycle is complete. And then it starts all over again…

We'd die without the water cycle

We depend on the water cycle. Our bodies need water. There is plenty in the sea. But we can't drink it, because it's too salty.

The water cycle turns **saltwater** into **fresh water**, which we can drink. It gets scattered as **rain**. The rain feeds rivers, and also sinks down into stores of water held in rock. The water we use at home is pumped from rivers and these underground stores.

Crops need fresh water too. The rain provides it. No rain, no crops. No crops, no food! So without the water cycle, we could not survive.

▲ *Borrowing from the water cycle.*

Rivers

How rainwater reaches the river

Rain makes rivers! Follow the numbers to see how:

Did you know?
- Right now, you are sitting above rock that's soaking in water.
- The water is called groundwater.

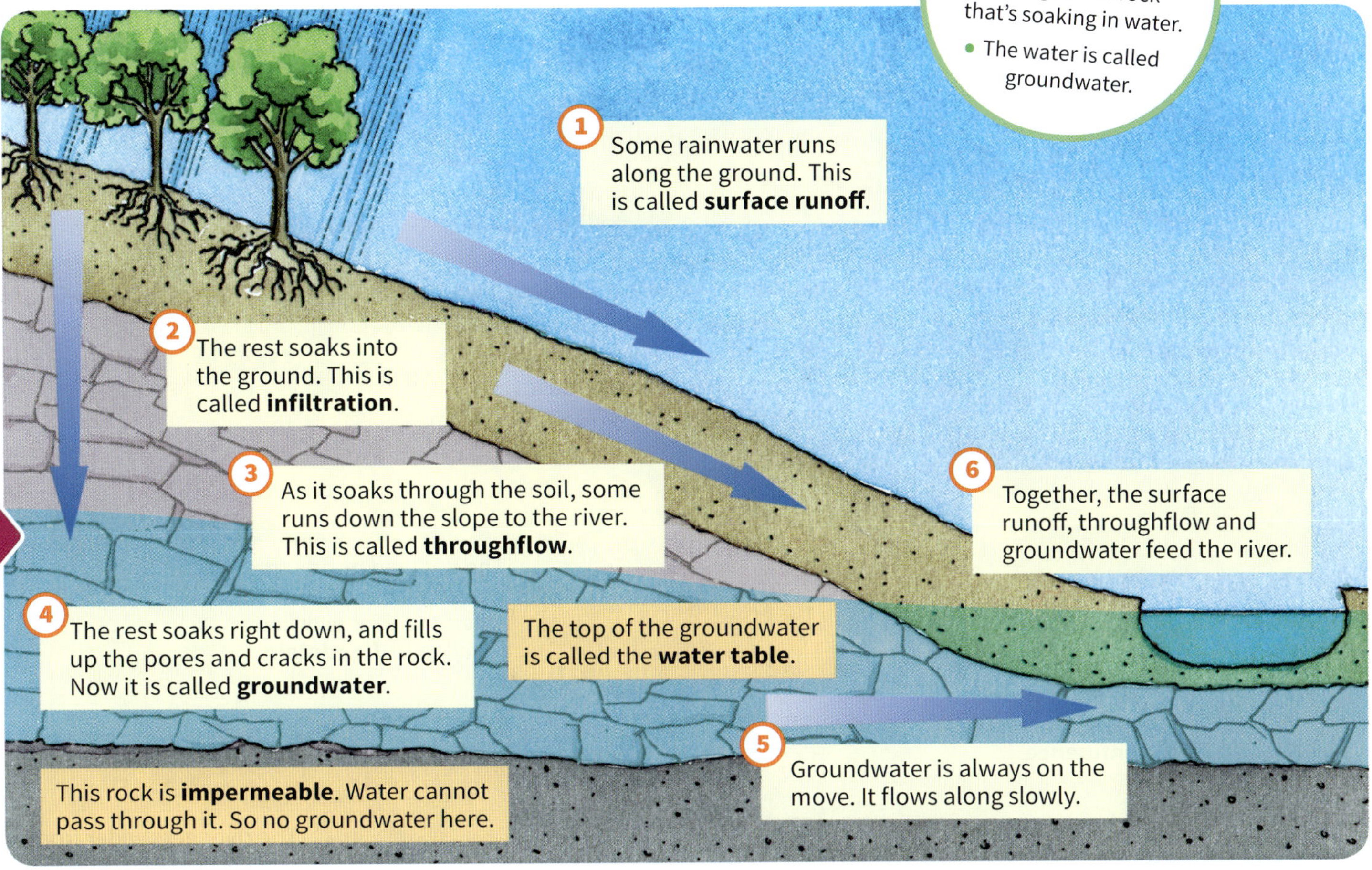

① Some rainwater runs along the ground. This is called **surface runoff**.

② The rest soaks into the ground. This is called **infiltration**.

③ As it soaks through the soil, some runs down the slope to the river. This is called **throughflow**.

④ The rest soaks right down, and fills up the pores and cracks in the rock. Now it is called **groundwater**.

⑤ Groundwater is always on the move. It flows along slowly.

⑥ Together, the surface runoff, throughflow and groundwater feed the river.

The top of the groundwater is called the **water table**.

This rock is **impermeable**. Water cannot pass through it. So no groundwater here.

Your turn

1. Make a larger copy of this flowchart for the water cycle.

   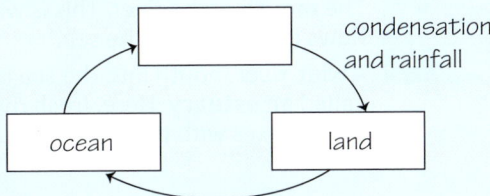

   Then add these labels in the correct places:
   rainwater feeds rivers   evaporation   atmosphere

2. a – g below are definitions of words used in this unit. For each definition, find the matching word in the unit. Then write out the word with its definition.
   a this water is held in rock, underground
   b the name for water in gas form
   c when water soaks down through the ground (i....)
   d a longer name for rainfall
   e the process that turns water into a gas (e....)
   f the process that turns water gas into water
   g does not let water pass through

3. Copy this diagram, to show how rainwater reaches a river. Add the missing labels, and a title.

   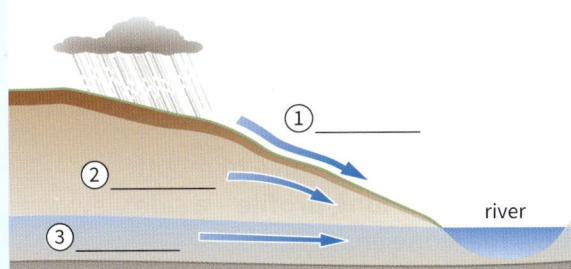

4. Use the big diagram above to help you explain why:
   a rain does not sink right down to the centre of Earth
   b a river can fill up very fast in very wet weather
   c a river will still have water in it, after weeks without rain

5. Two months ago, the water cycle stopped working! No more evaporation! No more rain!
   Write a radio report to describe how this has affected the UK, and how we're coping. Not more than 250 words.

81

5.3 A closer look at a river

Here you'll learn more about the course of a river – and take another look at the River Thames.

The river's journey

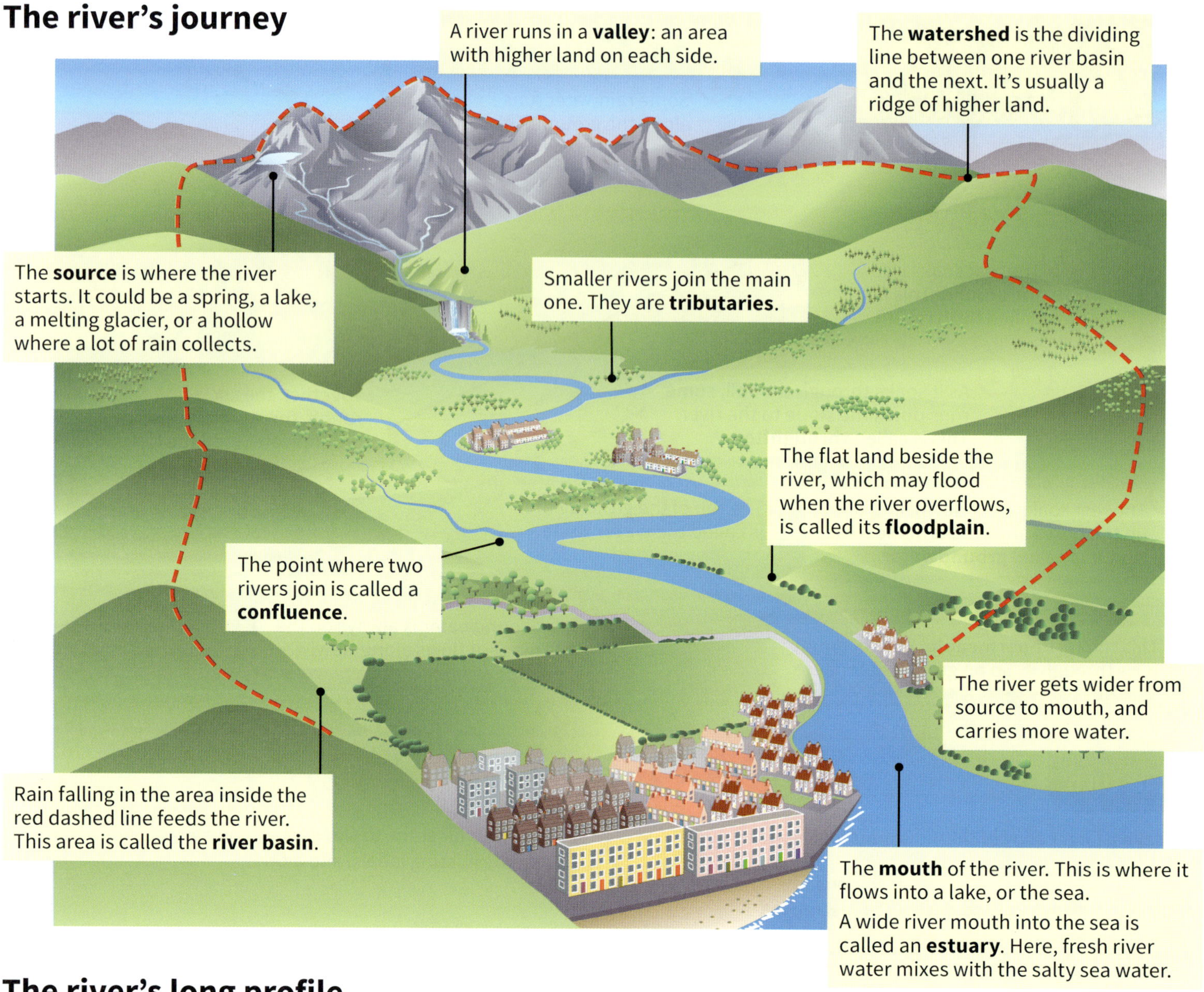

A river runs in a **valley**: an area with higher land on each side.
The **watershed** is the dividing line between one river basin and the next. It's usually a ridge of higher land.
The **source** is where the river starts. It could be a spring, a lake, a melting glacier, or a hollow where a lot of rain collects.
Smaller rivers join the main one. They are **tributaries**.
The flat land beside the river, which may flood when the river overflows, is called its **floodplain**.
The point where two rivers join is called a **confluence**.
The river gets wider from source to mouth, and carries more water.
Rain falling in the area inside the red dashed line feeds the river. This area is called the **river basin**.
The **mouth** of the river. This is where it flows into a lake, or the sea.
A wide river mouth into the sea is called an **estuary**. Here, fresh river water mixes with the salty sea water.

The river's long profile

A river flows downhill from source to mouth. This drawing shows its **long profile** – a side view, showing how the slope changes.

- The long profile curves down like a saucer.
- You can divide the river into three parts:
  - the **upper course**, where the slope is steepest
  - the **middle course**, where it is less steep
  - the **lower course**, where it flattens out.
- This river finally reaches **sea level**.

upper course
middle course
lower course
source of river
river
mouth of river
rock
sea

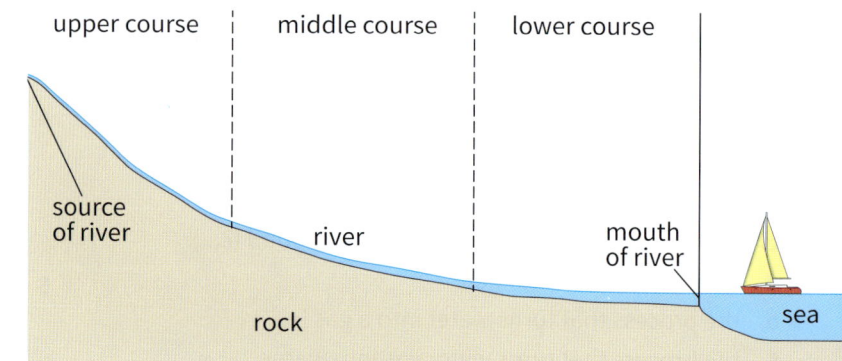

The river's channel

A river carves out a **channel** for itself.

This diagram shows the cross-section or **cross profile** of the channel (as if you'd sliced across it.)

The shape of the channel changes along the river, as you'll see later.

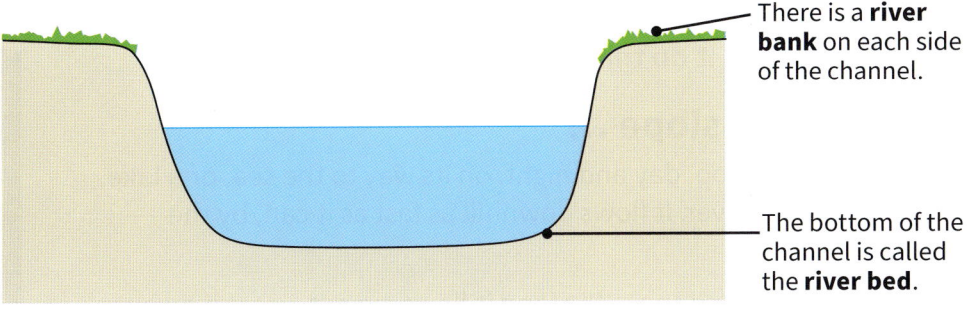

There is a **river bank** on each side of the channel.

The bottom of the channel is called the **river bed**.

Your turn

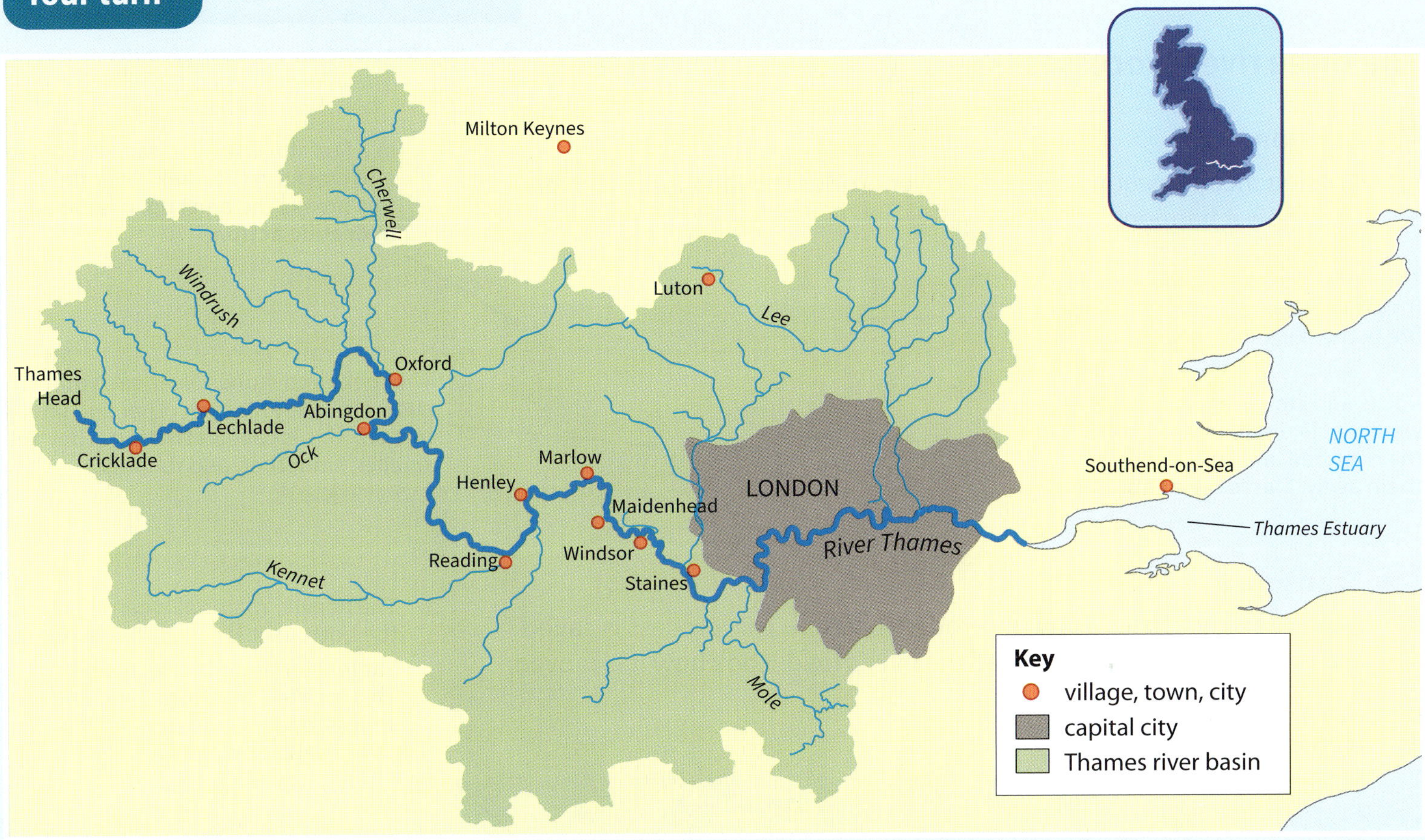

1 Terms A – F are about rivers. But they are jumbled up!
   - A *crouse*
   - B *tmuoh*
   - C *lavely*
   - D *tsyaeru*
   - E *ooldfianlp*
   - F *thareesdw*

   a First, unjumble each term.
   b Then define each one. Page 82 will help.

2 The map above shows the River Thames and its basin. Only some of the tributaries are named.
   a Name three tributaries of the River Thames.
   b Which tributary joins the Thames at Oxford?
   c Which one is shown in the photo and map on page 30?
   d Name the place at the confluence of the Thames and:
      i the Ock   ii the Kennet

3 Rain falling at Luton ends up in the Thames – at least 40 km away. Explain how this happens.

4 Will rain falling at Milton Keynes reach the Thames? Explain.

5 The first box below gives places on the Thames's journey.

Thames Estuary Windsor	12 m	
Lechlade Reading	110 m	66 m
Oxford Staines	0 m	29 m
Thames Head (source)	45 m	73 m

   a First, list the places in order, from the start of the journey.
   b Beside each place, write its height above sea level, in metres. Use the heights in the second box.
   c Now explain *why* the Thames flows downhill.

6 See if you can suggest reasons why the biggest settlement (town, city) along a river is often close to its mouth.

5.4 How do rivers shape the land?

A river changes the land it flows over. How does it do that? Find out here.

Off down the slope …

A river flows non-stop, day and night, on its way to the sea, or a lake, or to join another river. It flows downhill as fast as it can, by the easiest route.

As it flows, the river changes the land it flows over. It wears it away in some places by lifting material from it. It carries the material along. And then it drops it somewhere else. Let's look at those processes.

▲ Down the slope, by the easiest route.

The three river processes

1 Erosion

Erosion means wearing away. The river **erodes** the land it flows over. Look how it happens:

The water dissolves soluble minerals from the bed and banks. That helps to break them up. It is called **solution**.

Rocks and stones and sand in the water act like sandpaper. They scrape the river bed and banks and wear them away. This is called **abrasion**.

In a fast-flowing river, water is forced into cracks in the bank. Over time this breaks the bank up. It is called **hydraulic action**.

The rocks and stones wear each other away too. They hit each other and knock bits off. This is called **attrition**. They get smaller, smoother, and rounder.

2 Transport

Next, the river carries away the eroded material. This process is called **transport**. The material is called the river's **load**. Look how it is moved:

Dissolved material is carried along as a **solution**. You cannot see it.

Small light particles of rock and soil are carried along as a **suspension**. They make the water look cloudy or muddy.

The heavier material is carried along the bottom. It is called the **bedload**. Larger stones and rocks roll along. This is called **traction**. Sand and small stones bounce along. This is called **saltation**.

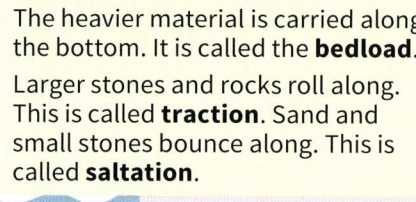

3 Deposition

Then, as the river reaches flatter land, it loses energy. So it drops or **deposits** its load. The deposited material is called **sediment**.

But dissolved material stays in the water. It will end up in the lake or the sea.

The biggest, heaviest, stones and pebbles are deposited first. Then smaller ones. And last, the smallest particles.

Rivers

What happens where?

This diagram shows how the balance between the processes changes, as you go down the river. The shape of the valley and channel change too.

> **Did you know?**
> - The Thames rises only 110 m above sea level …
> - … so its long profile is quite flat.

The river's long profile

upper course
Erosion is the main process in the upper course.

middle course
Both erosion and deposition occur here, in the river's middle course.

lower course
Deposition wins out in the lower course, as the river loses energy.

eroded material is transported downriver

sea

The cross profile of the river valley

floodplain

In the upper course, the river erodes downwards. The result is a **V-shaped valley**. (We'll look at this again later.)

Now the banks are being eroded, so the channel is getting wider. A **floodplain** forms where the river floods time after time.

In the lower course, the channel is even wider, and deeper. It holds a lot more water now. The floodplain has got wider too.

Your turn

1.
material is carried away	erosion
material is worn away	deposition
material is dropped	transport

   a The first list above shows the processes that go on in a river. Write them in the correct order.

   b Beside each, write the matching term from the second list.

2. Now look at photo **A** on page 84.

   a Which is the main river process going on here? Explain why you think so.

   b Describe the part played in this process by:
      i the water   ii the stones in the river
      Use terms shown in bold on page 84 in your answers.

3. Look at the river in photo **B**.

   a What is happening at X?

   b Is the river flowing quickly, or slowly, at X? Justify your answer.

   c Is B in the river's upper course? Explain your answer.

4. What is the difference between *saltation* and *traction*?

5. Explain *how* and *why* this changes, as you go down a river from source to mouth:

   a the volume of water in the river

   b the width of the river

   c the depth of the river

   d the amount of material dissolved in the water

B

5.5 Six landforms created by rivers

 Find out about six landforms a river may create on its journey.

The six landforms
This drawing shows six landforms created by rivers. The panels tell how they formed.

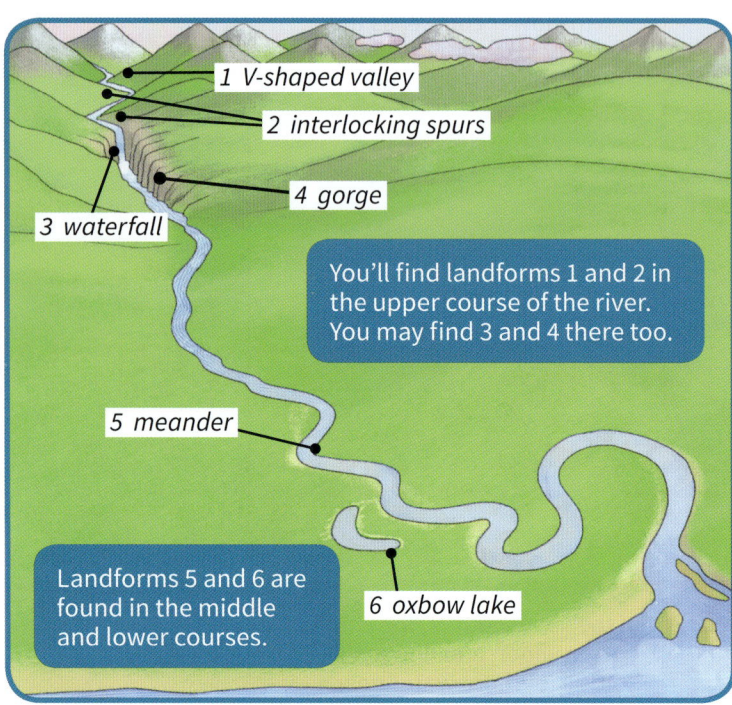

1 A V-shaped valley
In its upper course, a river erodes downwards, giving a steep valley. Over time, rain washes soil and stones down the sides. The valley becomes V-shaped.

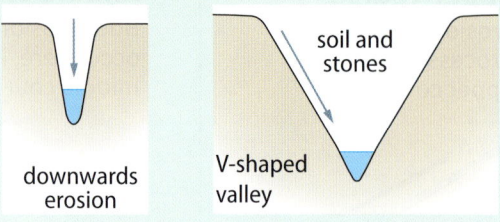

2 Interlocking spurs
A river takes the easiest route. It flows around rock that's hard to erode. The ridges or **spurs** of hard rock stick out, and interlock like fingers.

3 A waterfall
A waterfall is where water tumbles over a ledge of hard rock.

The waterfall forms where the hard rock meets soft rock. The soft rock gets eroded away.

4 From a waterfall to a gorge

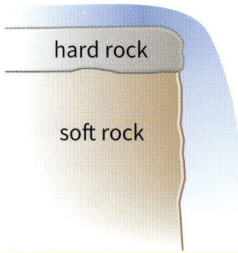

1 At a waterfall, the hard rock erodes very slowly. The soft rock below erodes much faster.

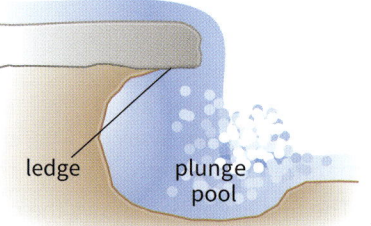

2 Erosion of the soft rock leaves a ledge of hard rock and a hollow called a **plunge pool**.

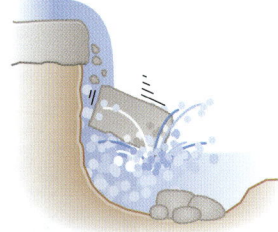

3 In time, the ledge falls into the plunge pool. The debris from it helps to speed up erosion.

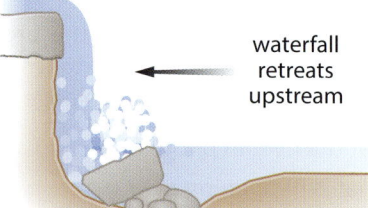

4 Steps 1–3 are repeated. The waterfall gradually retreats upstream, carving out a **gorge**.

Rivers

5 A meander

A **meander** is a big bend in a river. It starts as a slight bend. Look how it develops:

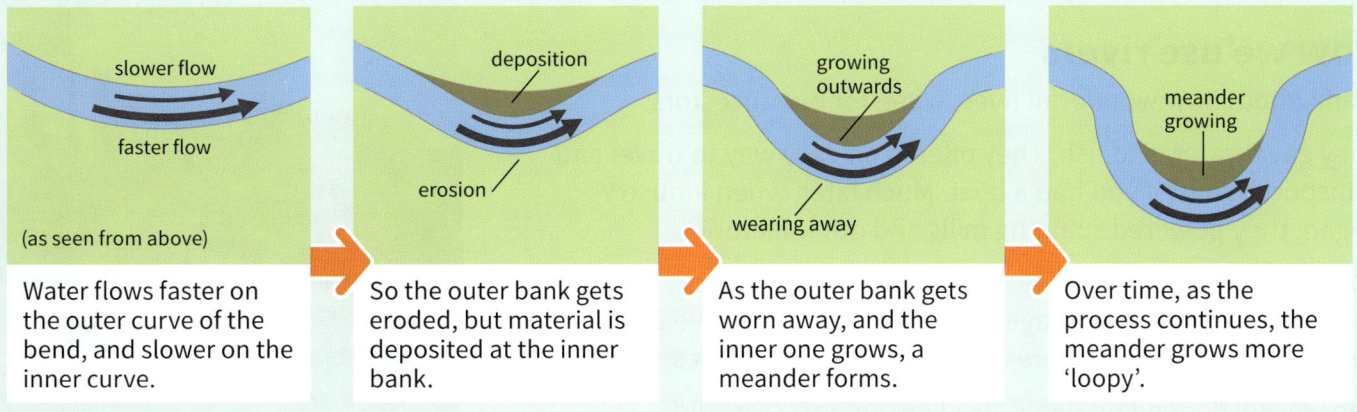

Water flows faster on the outer curve of the bend, and slower on the inner curve.	So the outer bank gets eroded, but material is deposited at the inner bank.	As the outer bank gets worn away, and the inner one grows, a meander forms.	Over time, as the process continues, the meander grows more 'loopy'.

6 An oxbow lake

An **oxbow lake** is a narrow U-shaped lake near a river. It's a meander that got cut off.

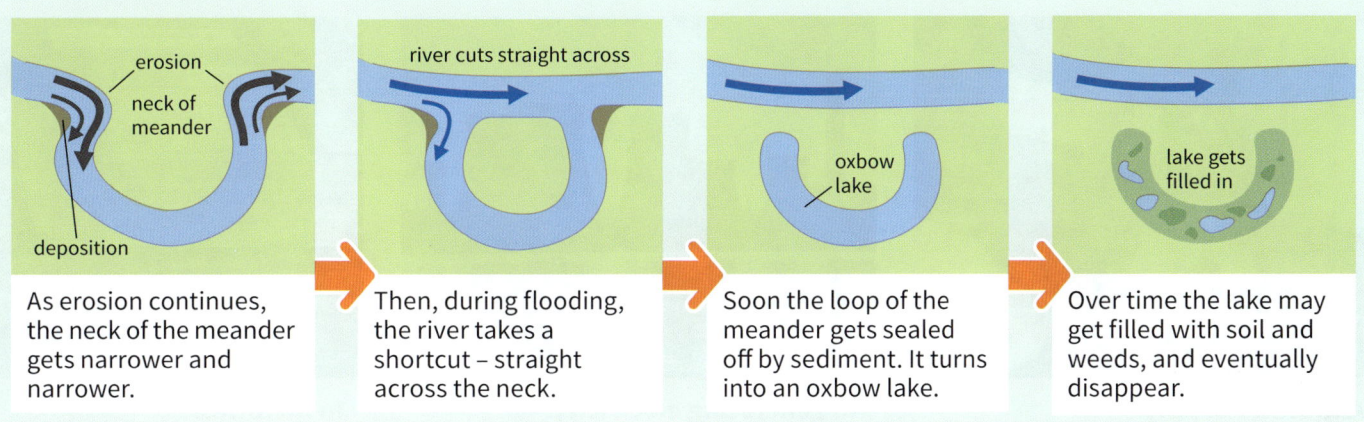

As erosion continues, the neck of the meander gets narrower and narrower.	Then, during flooding, the river takes a shortcut – straight across the neck.	Soon the loop of the meander gets sealed off by sediment. It turns into an oxbow lake.	Over time the lake may get filled with soil and weeds, and eventually disappear.

Your turn

1 Copy and complete this table, for the six landforms named in the first drawing on page 86.

Landform	Created by...
V-shaped valley	erosion

2 V-shaped valleys are found in the upper courses of rivers. Explain how erosion and rain create a V-shaped valley.

3 Draw and label a set of diagrams to show how a waterfall develops. Start with a diagram like this.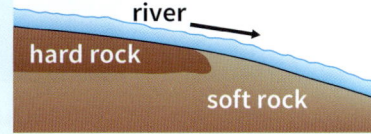

4 Most rivers have meanders.
   a Define the term *meander*.
   b Explain how a meander forms. (Use your own words, not copied from the text. You could give bullet points.)

5 The photo above shows meanders on a river.
   a Identify the process taking place:
      i at A   ii at B
   b Draw and label a sketch to show how this stretch of the river may look 150 years from now, after many floods.

5.6 How do we use rivers?

Find out how we use rivers – and harm river life. And see what you can do to help solve the problem.

How we use rivers

Think about how wonderful rivers were, for our ancestors.

They gave water, and fish. They offered an easy way to travel and transport things, if you had a boat. Much later, when industry began, they powered spinning mills and other factories.

So it's not surprising that we chose to live beside rivers. Over 70 **settlements** – villages, towns, and cities – have grown up along the Thames. There are others along its tributaries.

And we still depend on rivers, just like our ancestors did.

Making use of the River Thames

Let's see how people use the River Thames, as an example.

▲ *Windsor Castle, by the Thames. It's one of the royal homes. Henry VIII is buried in the chapel.* ("We are at home.")

As a water supply.
It's the main use of water from the Thames. Water is pumped from the river, cleaned up, and piped to millions of homes.

In generating electricity.
That's the second biggest use. In power stations, steam drives turbines. River water is then used to cool the tanks of steam.

In factories.
River water is used to wash materials, and cool equipment. This Ford factory in the Thames Estuary takes water from the river.

For farming.
The river flows through farmland on its journey. Some farmers use its water to irrigate crops, and as drinking water for their animals.

For transport.
Boats use the Thames all day long, for moving cargo and people. This barge carries cargo into London from a port in the estuary.

For leisure.
You can fish, canoe, swim, picnic on the bank, or walk along the river. (The Thames Path runs for 294 km, from the source to London.)

And now for the bad news …

We depend on rivers. So do the species that live in them: plants, insects, snails, toads, fish. And the animals that spend most of their time in the water: ducks, swans, herons, beavers, otters, water voles.

The bad news is … we are harming the river life in most of our rivers.

How we harm river life

1 **By polluting rivers.** Here are just three sources of pollution:

- **sewage**. Our waste water goes down the plughole or toilet. It flows to the sewage works. There it is cleaned up – and put back in the river! But quite often untreated sewage overflows into rivers at sewage works. (This may happen after very heavy rain.) It is toxic to river life.
- **waste water from street drains**. In some places this goes straight to a river, carrying things like engine oil and particles from car exhausts.
- **run-off from farms**. Fertilisers, pesticides, and animal waste can drain into rivers from nearby fields. They all harm river life.

2 **By pumping out too much water.** Billions of litres are pumped from rivers every day. (Much is then wasted!) So in dry weather the water level in the rivers falls. If it falls too low, fish and other river life will die.

▲ Feeding. But is it safe?

▲ Into the river it goes. But what's in it?

How *you* can look after river life

River life needs your help! Here are four simple rules.

- **Don't** throw litter into rivers. It lines the river bed, where many things live.
- **Don't** put things like face wipes and cotton buds down toilets. They block sewers, cause overflows, and end up in the river.
- **Don't** dump harmful stuff down sinks or drains. (Oil, paint, weed killer …) Chemicals from it may reach river life.
- **Don't** waste water. The more we use, the more must be pumped from rivers.

▲ Tiny water plants called **algae** growing over a river, helped by fertiliser from farms. When they die bacteria feed on them, using up oxygen from the water. So fish suffocate.

Your turn

1 This question is about how the River Thames is used. Choose:
 a four uses that are not likely to apply at Cricklade (page 78)
 b one use that is not likely to apply in central London
 c one use that is likely to apply to most rivers in the UK
 For each answer, explain your choice.

2 Over 13 million people live in the Thames drainage basin. Choose what you think is the most important use of the River Thames, for these people. Justify your choice.

3 a Why are power stations built along rivers, or at the coast?
 b Why are sewage treatment works built close to rivers?

4 The OS map on page 95 shows Purley-on-Thames, a suburb of Reading. It has several clues about past and present uses of the Thames. See how many you can find.

5 Look at the four rules above for helping river life. Choose the one you think is the most important. Explain your choice.

6 List at least five ways you could save water. (No funny ones!)

7 Think about this boy's opinion. Do you agree with it? Decide *Yes* or *No*. Then justify your answer.

We can't do without rivers, in the UK.

5.7 What's the Thames Estuary like?

The Thames Estuary is the final stage in the river's journey. Find out more here!

What is an estuary?

An estuary is a partly enclosed area of water, where a river meets the sea. Look at the Thames Estuary in **A**.

In an estuary, the fresh river water flows into the sea. And when the tide is rising, sea water flows up into the river channel. The two flows mix, giving **brackish** water. This is less salty than sea water.

Features of estuaries

The river and the tides deposit **sediment** in the estuary. So **salt marshes** and **mudflats** form. These are habitats for many plants and animals.

In the outer estuary, waves may deposit sand and pebbles, giving **beaches**.

The Thames Estuary

The Thames Estuary is important for wildlife. For example every year, hundreds of thousands of birds arrive for the winter, or to rest on their way further south.

It is important for humans too. There are towns along it. Hundreds of ships, barges, and boats sail on it every day. The **Port of London**, on its banks, is made up of over 70 shipping facilities. **London Gateway** and the **Port of Tilbury** are the biggest.

The estuary used to have lots of industries: cement making, steel making, oil refining and more. But many have shut down, leaving people without work.

Now there are plans to **regenerate** (revive) the area. With thousands of new homes, and centres for the creative industries – like music, computer gaming and film – where people can learn, and get jobs.

▲ The Thames Estuary. The numbers match the photos.

▲ A salt marsh at Leigh-on-Sea in the Thames Estuary. As the tide rises the channels fill with water. Fish breed in salt marshes.

▲ Brent geese on mudflats in the Thames Estuary. Every year they fly here from the Arctic, over 4000 km away, to spend the winter.

Did you know?
- The Thames Estuary was a target for enemy planes in World War II.
- Easy to pick out, and it leads to London.

Did you know?
- There are at least 1500 known shipwrecks in the Thames Estuary.

▶ Well hello! Seals are spotted all along the Thames Estuary, and porpoises too. Dolphins and whales sometimes swim in for a visit.

Rivers

▲ The roof of Cliffe Fort, built in the 1860s to guard the estuary. Now owned by a company that dredges gravel from the river bed.

▲ The Port of Tilbury. Ships carrying grain and paper and cars dock here. And container ships and cruise ships too.

▲ Danger! The masts of the shipwreck Richard Montgomery. *It has lain in the estuary since 1944, packed with explosives.*

▲ Thurrock Thameside Nature Park was once a landfill site. Then the rubbish was covered with soil. Lots of wildlife now!

▲ Southend-on-Sea, the seaside resort at the mouth of the estuary. It has the world's longest pleasure pier – 2.61 km!

Your turn

1 You met these terms in the unit. Unscramble them!
   a *fludamt*   b *tals harms*   c *sratuye*
   d *teesmind* e *trop*

2 Now define the terms you unscrambled in **1**. (Glossary?)

3 The Thames Estuary is a great location for a port. Explain how each factor below benefits the Port of London.
   a The estuary is near mainland Europe. (Map on page 141.)
   b The estuary is wide and deep.
   c London, with around 9 million people, is close by.

4 The land along the Thames Estuary is used in different ways.
   a Choose two photos that show a great contrast in land use.
   b Describe each of the photos you chose.
   c Explain why you chose that pair of photos.

5 a The Thames Estuary used to have lots of industries. Suggest two reasons why they were located there.
   b Explain why the area now needs *regeneration*.
   c Explain *how* these would help to regenerate the area:
      i world-class film studios   ii thousands of new homes

5.8 Floods!

What are floods? And what causes them? Find out here!

Did you know?
- The left bank of a river is the bank on your left as you look towards the mouth of the river.

What are floods?

Floods occur when water overflows the river's channel. Look:

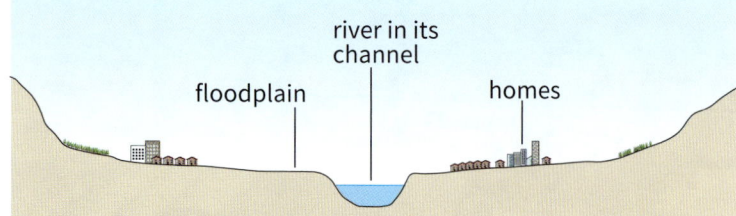

Here the river is flowing in its channel, as usual. Alongside the river is the **floodplain**, a fairly flat area that is likely to flood. There are people living on it.

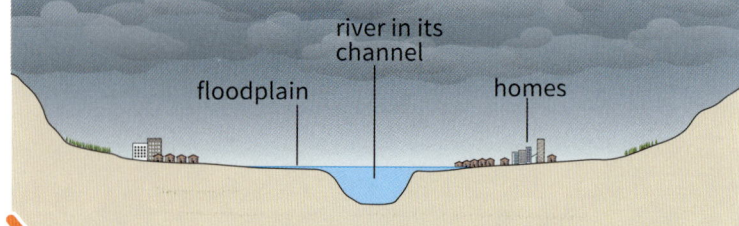

Now it has been raining heavily for weeks. The channel filled right up, and the water has overflowed. That means trouble for some of those people!

What causes floods?

Heavy rain is the main cause. It quickly finds its way to the river, as this drawing shows. The river's channel overflows.

When there is a lot of snow and it melts quickly, that can cause floods too.

Flash floods

A burst of very heavy rain can cause a sudden flood called a **flash flood**. This happens so fast that people get no warning. They can get trapped, and drown.

Adding to the flood risk

If rain can soak into the soil quickly, there is less chance of flooding. And the reverse is also true: anything that slows down or prevents infiltration will increase the flood risk. Look at the examples on the next page. Then try *Your turn*.

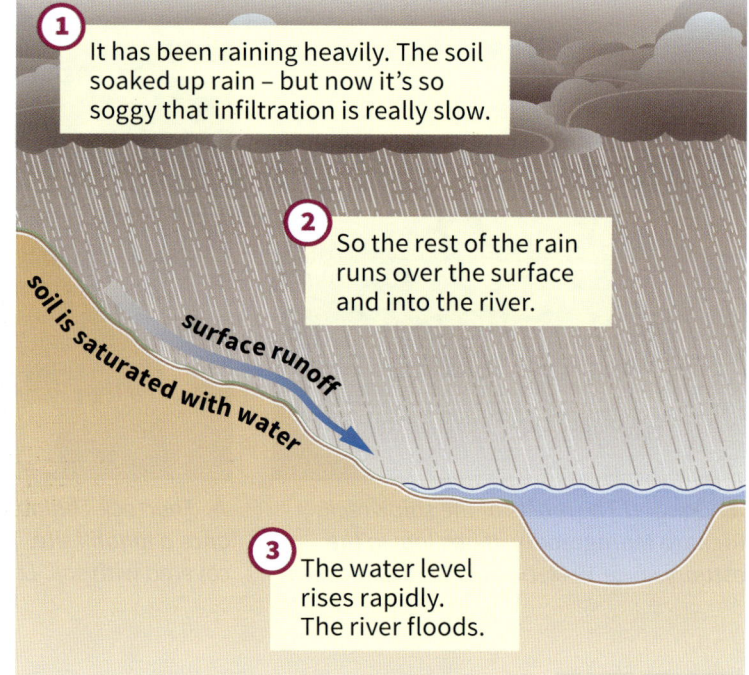

1. It has been raining heavily. The soil soaked up rain – but now it's so soggy that infiltration is really slow.
2. So the rest of the rain runs over the surface and into the river.
3. The water level rises rapidly. The river floods.

Your turn

1. The sentences below explain how a flood occurs. They are in the wrong order. Write them in the correct order.
   - The river fills up with water.
   - The ground gets soaked.
   - More rain runs over the ground and into the river.
   - Heavy rain falls for a long period.
   - The water rises over the banks.
   - Infiltration slows down.

2. **a** From the factors on page 93 that contribute to flooding, list: **i** the natural factors **ii** the human factors
   b Choose one natural factor and one human factor and explain *why* each increases the flood risk.
   c Which group of factors can we do something about?

3. **a** Look at the river on page 93. Explain why flooding would be less of a problem at X than at Y.
   b Suggest a way to stop floods reaching the homes at Y.

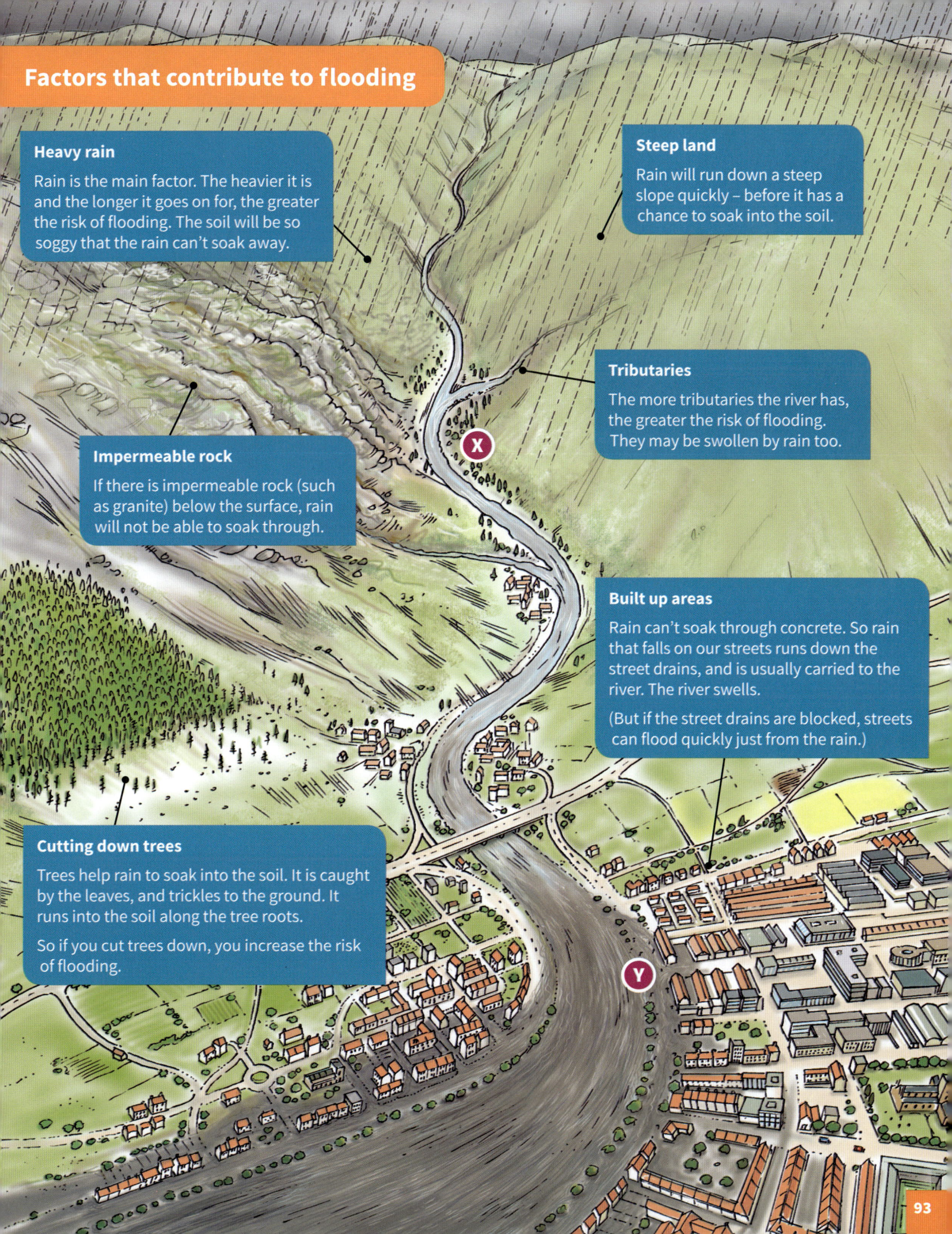

5.9 Flooding on the River Thames

Here you'll explore flooding on the River Thames, and some reasons for it.

Year after year …

Almost every year, there is flooding somewhere along the River Thames. Look at these photos.

A
▲ Richmond, London, 2018.

B
▲ Oxford, 2014.

C
▼ Datchet, 2014.

D
▼ Purley-on-Thames, 2012.

"Yes I DID get my socks wet."

E
▼ Wraysbury, 2014.

Your turn

The graph is for questions 1 – 4. It shows the monthly rainfall for one past year in the Thames basin. The red lines show the average rainfall for each month, calculated over the previous 30 years.

1. **a** How much rain fell in January 2012? Choose one:
      about 26 mm   about 42 mm   about 65 mm
   b Was January wetter or drier than average, that year?

2. There was a hosepipe ban in January – March of 2012, in the Thames basin. Explain why, using the graph and the facts in the blue panel.

3. The graph shows that April and June were much wetter than usual. But the Thames did not flood in those months. Suggest a reason. (To do with soil?)

4. **a** The ground in the Thames basin was really dry at the end of March. But by the end of October it could not hold any more water. It was saturated. Explain why.
   b The Thames began to flood in November, and flooding continued in December. Explain why.

5. Photo **D** shows flooding in Purley-on-Thames, a suburb of Reading, in 2012. Purley is shown on the OS map below.
   a Purley lies in the floodplain of the Thames. How can you tell this from the OS map? (Hint: contour lines!)
   b Which is more likely to flood: a house at 668764 or a house at 671768? Explain your answer.

6. You want to prevent flooding of the houses along the river in square 6676. What will you do? Outline your plan.

7. Choose any photo **A** - **E**. What's it like there? What damage has been done? Write a report for radio. (At least 10 lines.)

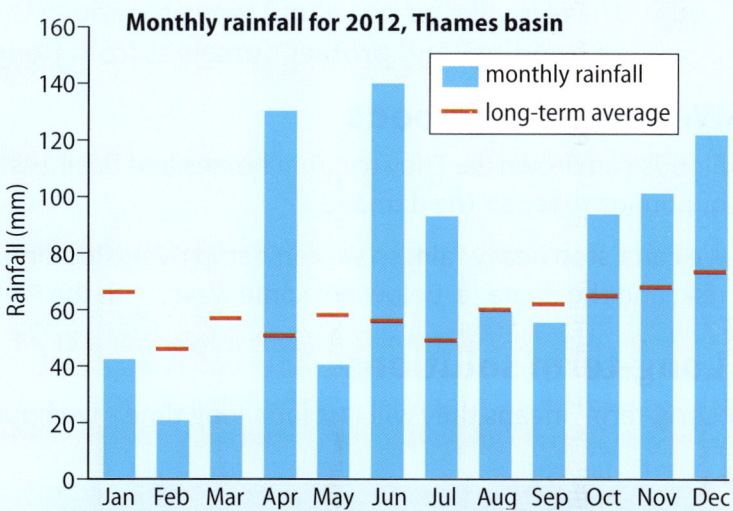

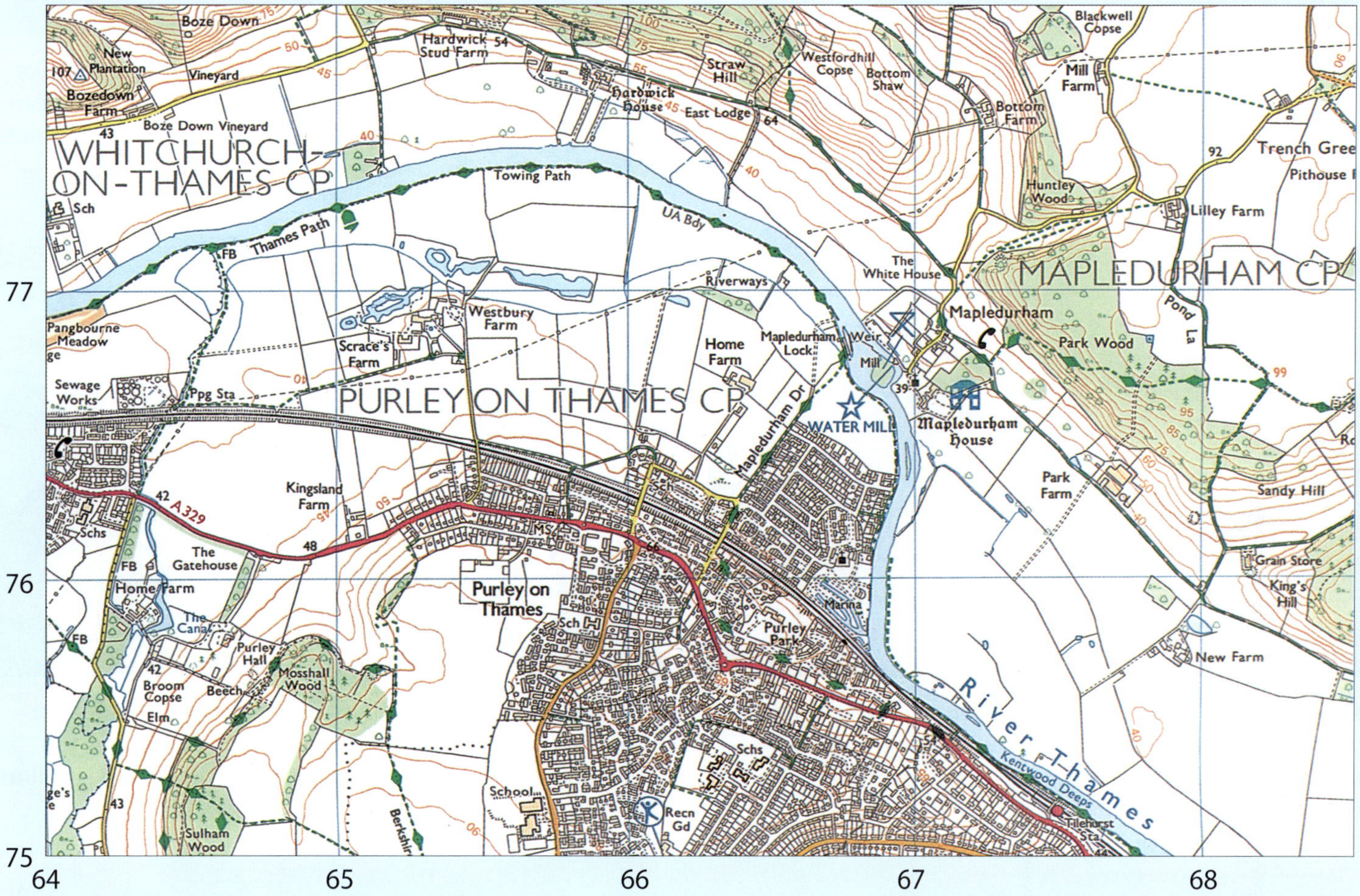

The winter of 2011 was unusually dry, in the Thames basin.

By the start of 2012, groundwater levels were low.

Scale 1 : 25 000

© Crown copyright

5.10 Can we protect ourselves from floods?

Here you'll find out about ways to reduce the risk of flooding, and protect ourselves from floods.

We can't stop floods

Floods can drown us. They can ruin homes and businesses. It can cost billions of pounds to repair the damage.

We can't stop heavy rain, so we can't stop rivers flooding. But we can reduce the risk, and the damage. Below are some ways, with the River Thames as example.

Long-term solutions

'Long-term' means they will last for a long time – we hope!

Did you know?
- The world's most deadly floods were in China in 1931, when the Yangtze and Huai rivers flooded.
- It's thought that over 3 million people drowned.

A **Build embankments (high banks).** These embankments along the Thames in London were first built to hold an underground sewage system, and Tube lines. But later, the walls were made higher as flood protection.

B **Dig new river channels.** The Jubilee River above looks natural – but it's not! It was dug out to divert water from the Thames, in order to prevent flooding at Maidenhead, Windsor, and Eton. There are plans for others too.

C **Take care where you build.** Local councils now weigh up the flood risk, before they allow new buildings in the Thames floodplain. They may refuse planning permission. You'd have to find another site.

D **Let nature help.** Allow land along the river to soak up flood water, as nature intended. Plant more trees too. This common land beside the Thames in Oxford is called Port Meadow. It regularly floods. The horses walk away!

Rivers

Short-term solutions

When we know floods are on the way, here are some things we can do.

Put up portable flood barriers. This shows them being fitted in Oxford, next to the river, after a flood warning. They'll be taken down later and stored away.

Put anti-flood shutters on homes. You can buy metal shutters like these to stop water coming in through doors and windows. (Or else try sandbags.)

The Thames Barrier

London can also be flooded from the sea. So it has special flood protection: the **Thames Barrier**.

This barrier has a set of giant steel gates which lie on the river bed. They are raised when there's a risk that high water levels on the Thames will meet high tides coming in from the sea. Their job is to shut the sea water out.

Who decides?

The **Environment Agency** works with local councils in England, to decide what to do about flooding. It gets a grant from the government to install anti-flood structures.

It also monitors water levels in rivers, and gives out flood warnings for areas at risk.

▲ The Thames Barrier. These piers hold machinery for raising big steel gates which lie flat on the river bed. The gates swing up to close the gaps between the piers.

Your turn

1 This is about the long-term solutions on page 96. You can answer using their picture labels, **A – D**.
   a Which solutions aim to stop flood water overflowing?
   b Which one aims to keep us away from floods?
   c Which two are likely to cost most?
   d Which two do you prefer? Explain why.

2 Describe how these protect people from flooding. (You could draw simple diagrams, with notes.)
   a embankments   b digging a new river channel
   c setting aside fields to take flood water

3 You are in charge of building a new town near a river. Which solution(s) will you choose, to reduce the risk of flooding? Explain your choice. (If you wish, you can answer by sketching a map of the imaginary town and river, and adding notes.)

4 The solutions in **E** and **F** are called *short-term*. Why?

5 Do you agree with this person's proposal? Write a thoughtful reply.

6 Now see if you can design a flood-proof home. Draw sketches!

5 Rivers

Let's see how much you've learned about rivers ...

check ✓

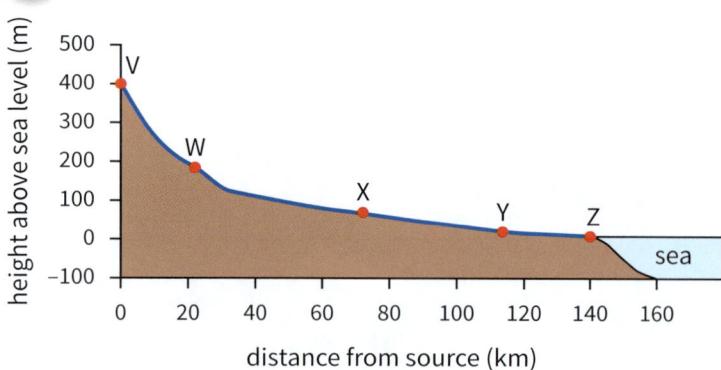
A

Questions 1 – 4 are about diagram A above.

1 A shows the long profile for a river which flows to the sea. Letters **V – Z** represent places along the river.
   a Which letter represents the source of the river?
   b Which letter represents the mouth of the river?
   c How long is this river? (Careful!)
   d How high is the source of the river above sea level?
   e Describe the shape of the river's long profile.

2 B shows a cross profile of the river valley, for the river in **A**. Match **B** to one of these two locations on the river:
   a W   b Y
   Justify your choice.

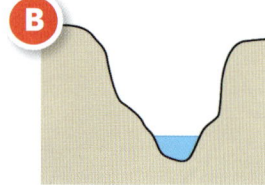
B

3 a At which place in **A** is there likely to be more deposition?
      i at W   ii at Y
      Justify your answer.
   b There are stones all along the river bed. Where are they likely to be smaller and rounder?
      i at W   ii at Y
      Explain your answer.
   c The water level at **Z** rises and falls continually. Why?
   d There are mudflats at **Z**. Explain why.

4 Here are some facts about the river in **A**.
   a The river is 6 m wide at **W**, and 12 m wide at **Y**.
   b The river has about five times more water at **Y** than at **W**.
   For each fact, suggest an explanation.

5 Turn to the OS map on page 35. It shows the River Coquet.
   a Identify the river feature shown in square 2405.
   b The River Coquet has a small estuary. Define *estuary*.
   c i Identify two natural habitats on the OS map that are usually found in estuaries. (See the key on page 34.)
      ii Give a grid reference for a square with both habitats.

6 **C** shows the Clutha River in New Zealand. Look at the feature marked **X**. Much of it is covered in green algae (water plants).

C

   a Identify the feature marked **X**.
   b Draw labelled diagrams to show how this feature formed.
   c i This feature is changing. In what way?
      ii Describe how you'd expect it to look 100 years from now.

7 **D** shows flooding at Mirfield, a town on the River Calder in the north of England.

D

   a Suggest the main reason why the River Calder flooded
   b Look at the building marked **Y**. A building might not be permitted on that site today. Explain why.
   c There's flooding on the road near **Y** (where the white van is). Suggest one way to protect this road from floods.
   d Floods have an *economic impact*. That means they cost people money – to fix things, or because they lose business. Suggest two examples of their economic impact on the owner / driver of the white van.

8 *We need to look after our rivers.*
   Discuss this topic. Write at least half a page. Think about the importance of rivers to us, and the impacts of pollution, and our heavy use of water, on the health of rivers. And check *discuss* on page 18.

6 Africa

6.1 What and where is Africa?

 Here you'll compare Africa with other continents – and think about your mental images of Africa.

Africa: a continent

Africa is not a country. It is one of the world's seven **continents**. Look at this map.

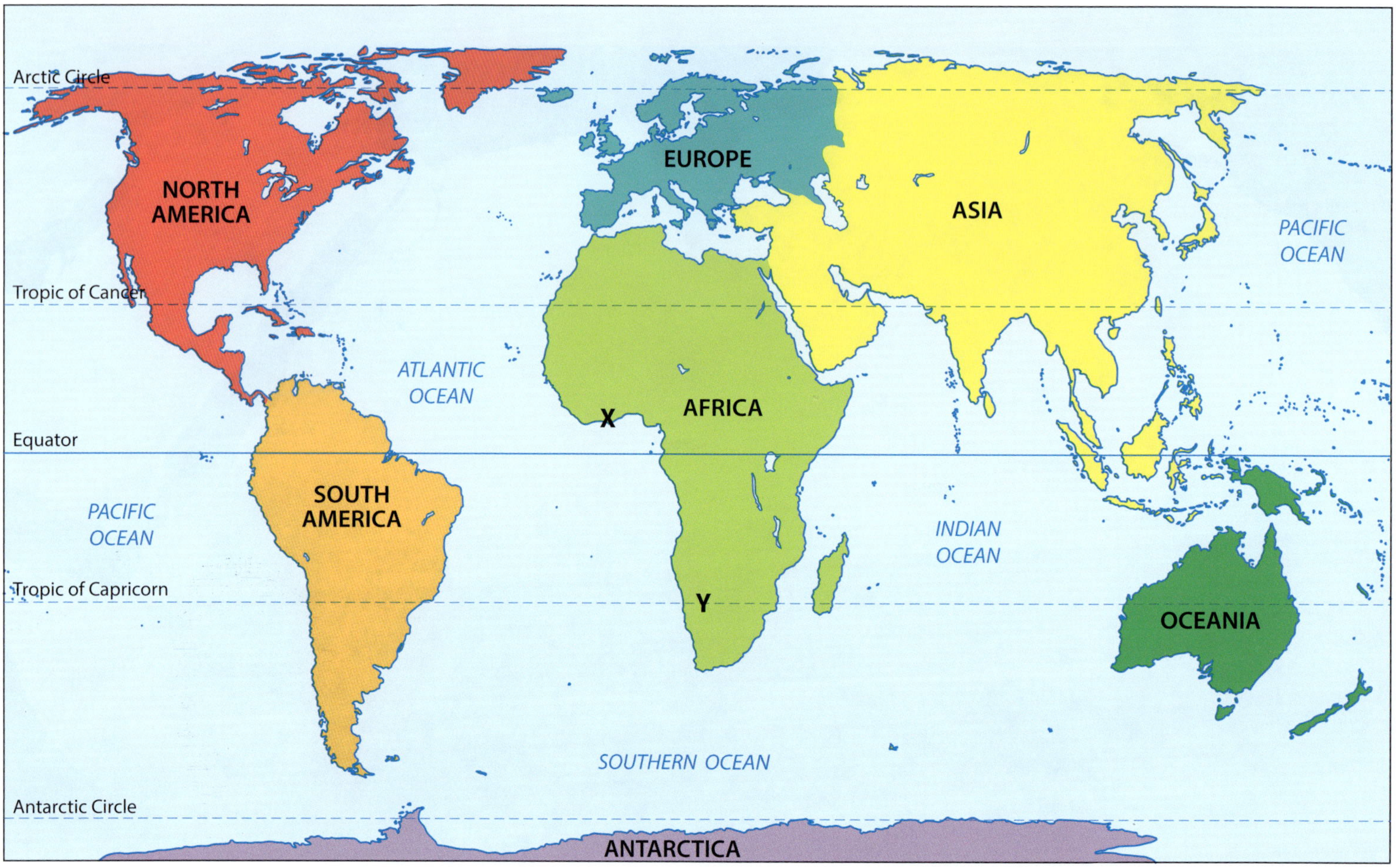

Did you know?
- At their closest points, Africa is only 14 km from Europe.

Note how the Equator runs across the middle of Africa. Now look at the lines for the tropics. Most of Africa lies within the tropics – between those two lines.

Compare it with the others

Africa is the world's second-largest continent, for both area and population. Look at these tables.

Did you know?
- Most of the world's best long-distance runners are African.

The continents by land area

Continent	millions of square km
Asia	44.6
Africa	30.1
North America	24.5
South America	17.8
Antarctica	13.2
Europe	9.9
Oceania	8.1

The continents by population

Continent	millions of people
Asia	4545 (or 4.545 billion)
Africa	1287 (or 1.287 billion)
Europe	743
North America	588
South America	428
Oceania	41
Antarctica	people only visit

Africa

Your turn

1 Lots of people think Africa is a country. It's not. It is a continent.
   a Explain the difference between a *country* and a *continent*. (Glossary?)
   b Now, a challenge. See how many African countries you can list, without looking at a map. (There are 54!)
   c Swop the list from **b** with your partner, to score it. Turn to the map on page 106. Give 1 mark for each correct name. Subtract 1 mark for each wrong one. How did you do?

2 Look. This graph compares the **areas** of the continents.

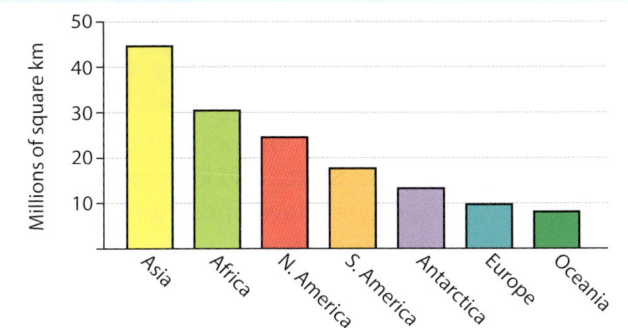

Using the graph, decide whether each statement below is true, or false. (You can check the table on page 100 too.)
   A All of North America would fit into Africa.
   B South America and Europe together would fit into Africa.
   C Africa is about three times the size of Europe.

3 This graph compares the **populations** of the continents.

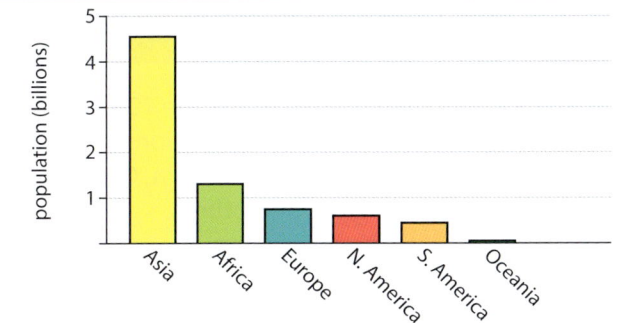

True or false? (Check the table on page 100 too.)
   A There are more people in Africa than in Europe.
   B Africa has more people than North and South America combined.
   C More people live inside Asia than outside Asia.

4 The **population density** of a place is the average number of people living there per square kilometre.
The population densities for the continents, rounded off, are:

Africa, 40   Antarctica, 0   Asia, 99   Europe, 75
North America, 24   South America, 24   Oceania, 5

See if you can find a good way to display this data. (A bar chart? A pictogram?)

5 True or false? Your answer for question **4** will help you decide.
   A Africa is the most crowded of the inhabited continents.
   B Europe has nearly twice as many people per square km as Africa does, on average.
   C Africa has more people per square km than either North or South America, on average.

6 This photo was taken at the place marked **X** on the map.

   a There is an ocean in the photo. What is its name?
   b Name the ocean off the east coast of Africa.
   c At **X**, the days are warm all year round. See if you can explain why. (Look at the blue lines across the map?)

7 We all have **mental images** of Africa (pictures in our minds).
   a From where did you pick up your mental images of Africa?

   b The city above is Windhoek. It is at **Y** on the map. Does this photo fit with your mental images of Africa? Explain.

8 Some people think Africa is full of poor people, with no hope. This is a **stereotyped** view.
   a What does *stereotyped* mean? (Glossary?)
   b Suggest some things you could do to make sure that your views of Africa are not stereotyped.

9 Now it's time to start a big spider map, as your summary for Africa. Use a double page. Mark in facts you know already. You could group them under headings such as:
Where is Africa?
You can add more to your spider map later.

6.2 A little history

 For a time, most of Africa was 'owned' by Europe. Find out more here.

Africa: our cradle

We are all linked to Africa. Experts think that we (*Homo sapiens*) evolved there. Fossils over 310 000 years old have been found in Morocco.

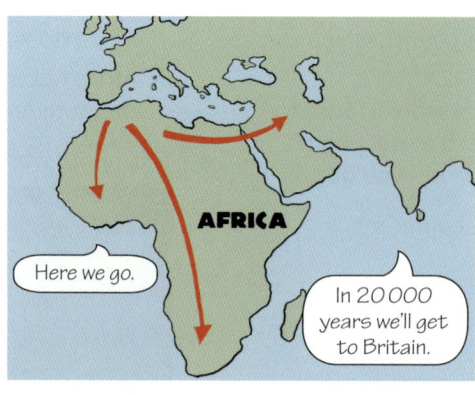

At some point – perhaps 100 000 years ago – we began to migrate. Within Africa, and out of Africa to the rest of the world.

Over time, the population of Africa grew. States and kingdoms and empires developed. Some were very wealthy.

Africa had early trading links with Asia. By 800 CE, traders were already sailing to East Africa from the Arabian Peninsula, and India.

Europeans began to explore Africa in the 15th century. The Portuguese were first, sailing down the west coast and up the east coast.

From the 15th to the 19th centuries, other European nations followed: Britain, France, Spain, Germany, Belgium and others.

At first, the Europeans were keen on trading. They wanted things like gold, ivory, rubber, timber, cocoa, and fruits. And slaves.

The European nations became greedier – and fierce rivals in their trade with Africa. This often led to violence.

So, to avoid conflict, they met in Berlin in 1884 to carve Africa up between them. The meeting is known as the **Berlin Conference**.

Africa

After agreeing on new country borders, the European nations set up colonies. By 1914, almost 90% of Africa had been colonised.

But one by one the colonies rebelled, and fought for independence. The last to gain independence was Zimbabwe (from the UK) in 1980.

Today Africa – which once had its own states and kingdoms and empires – is made up of 54 independent countries, shaped by the colonisers.

The impact of colonisation

When the Europeans came to Africa, the world was more equal than it is now. Parts of Africa were rich, through trade with other African countries and the Arab world.

Today the world is very unequal – and most of the poorest countries are in Africa. Colonisation is at least partly to blame.

- The European colonisers took away resources (like gold, timber, rubber). So they got richer. Africa did not.
- European slave traders took at least 10 million Africans to work in plantations in North America and the Caribbean. This made European nations richer. Africa lost.
- When the country borders were drawn, different ethnic groups were forced together. This often caused conflict – and still does. Conflict holds a country back.
- At independence, the colonies had to start governing themselves. They did not have all the systems in place for this. Many struggled to cope – and some still do.

Who colonised Africa?

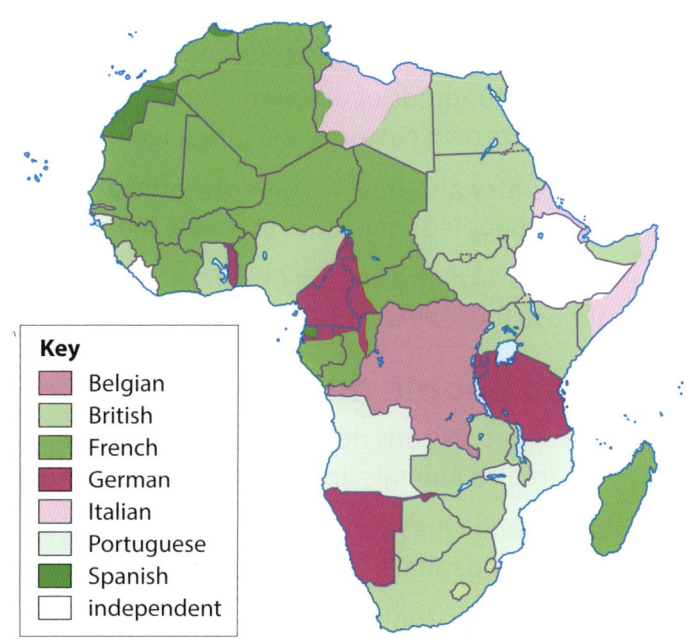

Your turn

1. We all share one link with Africa. What is it?
2. Define these terms. (Glossary?)
   a colony   b coloniser   c independence
3. The European colonisers wanted raw materials from Africa.
   a Define *raw material*.
   b Name one raw material they obtained from Africa.
   c Explain why the colonisation of Africa helped to make the European colonisers wealthy.
4. The sailing ship played a big part in the colonisation of Africa. See if you can explain why.
5. Suggest two reasons why the Portuguese were the first Europeans to explore the African coastline. (Page 141.)
6. Look at the map above. Two European countries between them colonised about two-thirds of Africa. Which two?
7. See how many African countries you can name, that were British colonies. (The map on page 106 will help.)
8. Suggest one way in which this harmed Africa:
   a the slave trade
   b different ethnic groups, with different customs and languages, being forced together in a new country

6.3 What's Africa like today?

Africa is a big, vibrant, exciting continent. This unit will give you an overview of Africa today.

Africa's countries
- Africa is the world's second largest continent – in both area and population. (Asia is first.)
- Its 54 countries are all different. Some are mostly desert. Many are tropical.
- 23 of them are smaller than the UK, in area.
- Only 4 of them have more people than the UK!

Africa's people
- Africa has around 1.3 billion people. (That's 1 300 000 000.) The number is growing fast.
- It's a **young** population. Over half of the population (51%) is under 20 years old. (Around a quarter of the UK's population is under 20.)
- There are hundreds of different ethnic groups across the 54 countries. Each has its own customs and language.
- Each country also has two or more 'official' languages, taught in school. So most people speak more than one language.
- Around 49% of Africans are Christian, and 43% are Muslim. (North Africa is almost 100% Muslim.) Other religions are practised too.

What do people do?
- Over 60% of Africans depend on farming for a living. They grow crops and / or raise animals. In some countries, it's up to 90%.
- In towns and cities you'll find people working in shops and schools and banks and hospitals, and driving buses – the same as people here.
- Most countries don't have a lot of industry – yet. But that's changing.

▲ *A secondary school class in Kenya. If you lined up everyone in Africa in order of age, the person in the middle would be 19.*

▲ *A dad and his daughter in Egypt.*

▼ *A rural scene in Uganda.*

▼ *Yachts in a marina in Durban, a city in South Africa.*

Africa's natural wealth

Africa has many **natural resources** which it can use to earn money.

- It has large deposits of metal ores, including copper, cobalt, uranium, iron – and gold. It has diamonds too.
- It has 8% of the world's known oil and gas reserves – and may find more.
- But natural resources are never spread evenly. Only some countries have diamond mines, for example.
- Most of the big companies who extract the ores, diamonds, oil and gas are in fact foreign companies. Much of their profits leave Africa.
- Many countries grow **cash crops** like cotton, tea, coffee, rubber, fruits, and flowers, for export.
- Most of the exported materials are **processed** in the countries that buy them. For example copper is turned into electric cables. This **adds value**. Cables can be sold for a lot more than the copper itself.

▲ A diamond mine in Sierra Leone. Buckets of mud are passed up the slope. At the top people sieve the mud, and pick out the diamonds.

So ... is Africa rich, or poor?

- In spite of its natural resources, Africa is the poorest continent.
- Almost half of its population lives in **extreme poverty**. They have less than about £1.50 a day to live on. For everything.
- But not everyone is poor. Some countries are much better off than others, and even the poorest countries have some wealthy people.
- And things are improving. The **economies** of several African countries are now among the fastest growing in the world.
- All African countries are working to banish extreme poverty by 2030. This is part of the **Global Goals for 2030** agreed by world leaders.

Did you know?
- China is Africa's top trading partner.
- There are over 1 million Chinese living in Africa.

Your turn

1. The population of Africa is about ... how many?
2. Over half of Africa's population is aged under 20. Think about this. Then suggest:
   a one advantage   b one disadvantage
   of having such a young population.
3. Africa's population is expected to double in the next 35 years. What problems might this fast growth cause? (Think about things people need ... like food, shelter, water.)
4. English is the official language in Ghana in Africa, but French is the official language in its neighbour, Côte d'Ivoire. Suggest a reason for this difference.
5. Define *natural resource*. (Glossary?)
6. Turning copper into electric cables *adds value*.
   a Explain the term in italics.
   b Africa would benefit if it made cables and other useful items from copper, and exported these. Explain why.
   c Suggest one reason why Africa does not yet make lots of useful items from copper, for export. (Factories?)
7. *Extreme poverty* is when people live on less than $1.90 a day. That's about £1.50 a day in pounds – or about £550 a year.
   a Name one thing that costs about £550.
   b Now list all the things a person anywhere in the world needs, to survive for a year. (Food, shelter, fuel ...)
8. Choose the photo that interests you most in this unit, and explain why you chose it.

6.4 The countries in Africa

Africa has 54 countries. Find out more about them here.

The countries and their capitals

Look at all these countries! Can you spot any you have heard of?

Compare!

Key
MALI country names are shown like this
■ capital cities

Did you know?
• Only Ethiopia and Liberia were never truly colonised.

Africa

◀ Cairo, Africa's second largest city. (Lagos is first.) The city has over 12 million people. That river is the Nile.

Did you know?
- Africa's top football competition is the Africa Cup of Nations …
- … played every second (odd-numbered) year.

Your turn

1 Identify this African country from map **A**.
   a It's on the east coast, and its name begins with K.
   b It is tiny, and completely surrounded by South Africa.
   c It is a big island, larger than the British Isles.
   d It is just north of Nigeria, and a bit larger.
   e It is small and thin; its name starts with T and has 4 letters.

2 Now see how many countries you can find, beginning with:
   a M   b Z   c L   d S

3 When the European colonisers met in 1884 to carve up Africa, they drew new country borders on a map – sometimes using a ruler!

   Look at map **A**. Which countries have some borders that appear to have been drawn with a ruler? Name at least seven.

4 When the colonisers bargained over Africa, they had these issues in mind:
   A natural resources   B climate
   C access to the coast

   Suggest reasons why each issue was important to them.

5 Below are some capital cities, from the map.
   For each, see how fast you can find the matching country.
   Then write down the country and its capital.
   a Addis Ababa   b Lusaka
   c Nairobi   d Tripoli
   e Nouakchott   f Kampala

6 Africa has different regions. Look at map **C**.

   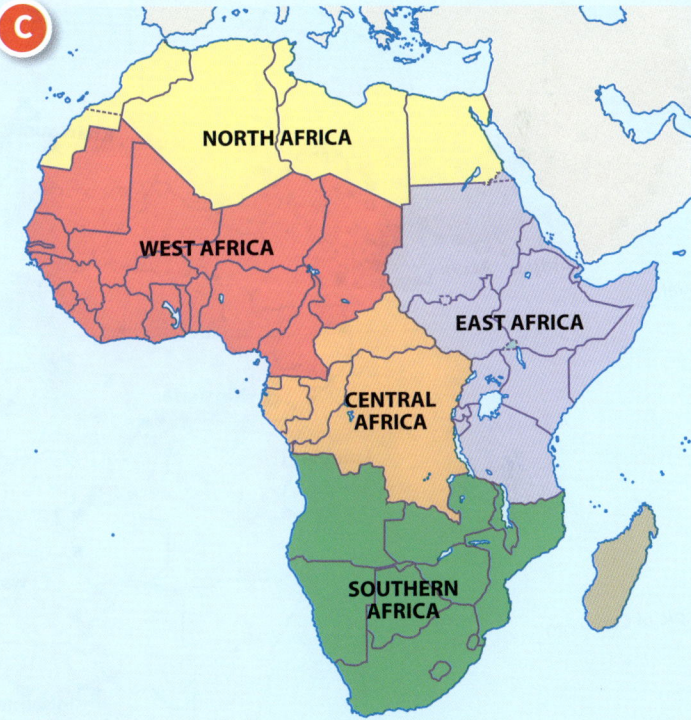

   Name countries in each region, as follows:
   a three in Central Africa   b four in North Africa
   c three in East Africa   d four in West Africa
   e four in Southern Africa

7 Now look at Cairo, in photo **B**.
   a It is the capital of … ?
   b Find out as much as you can about Cairo from the photo. About buildings, open spaces, trees, the river, air quality.
   c Use what you observed to write a paragraph about Cairo.

6.5 Population distribution in Africa

 This unit is about where people live, in Africa. You will explore a table of data, and a map.

How many people, and where?

Africa is home to about 1.3 billion people. Look at the list on the right. It gives the **population** of each country in millions. (1 million = 1 000 000.)

People are not spread out evenly. Some places are more crowded. The map below shows the **population density**. The deeper the shade, the more people there are, in that area.

A Map of Africa showing population density and major cities.

Key

Population density — people per square kilometre
- over 100
- 10–100
- 1–10
- under 1
- country border

Major cities — population in millions
- over 3
- 1–3
- 0.5–1
- 0.1–0.5

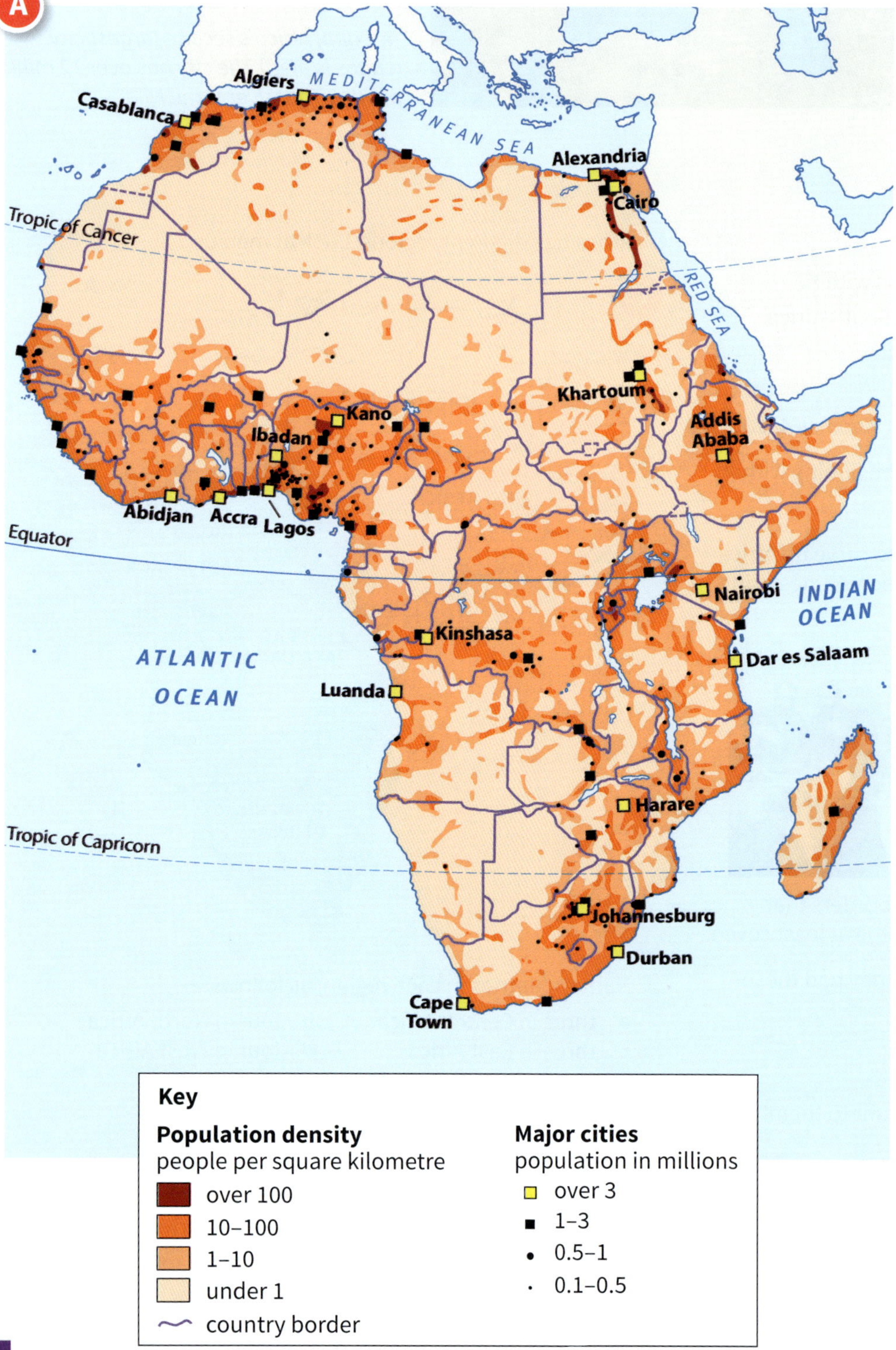

B

Country	Population (millions)
Nigeria	195.9
Ethiopia	107.5
Egypt	99.4
Dem. Rep. of Congo	84.0
Tanzania	59.1
South Africa	57.4
Kenya	51.0
Uganda	44.3
Algeria	42.0
Sudan	41.5
Morocco	36.2
Angola	30.8
Mozambique	30.5
Ghana	29.5
Madagascar	26.3
Côte d'Ivoire	24.9
Cameroon	24.7
Niger	22.3
Burkina Faso	19.8
Malawi	19.2
Mali	19.1
Zambia	17.6
Zimbabwe	16.9
Senegal	16.3
Chad	15.4
Somalia	15.2
Guinea	13.1
South Sudan	12.9
Rwanda	12.5
Tunisia	11.7
Benin	11.5
Burundi	11.2
Togo	8.0
Sierra Leone	7.7
Libya	6.5
Congo	5.4
Eritrea	5.2
Liberia	4.9
Central African Republic	4.7
Mauritania	4.5
Namibia	2.6
Botswana	2.3
Lesotho	2.3
Gambia	2.2
Gabon	2.1
Guinea-Bissau	1.9
Swaziland	1.4
Equatorial Guinea	1.3
Mauritius	1.3
Djibouti	1.0
Comoros	0.8
Cape Verde	0.6
São Tomé and Príncipe	0.2
Seychelles	0.1

Africa

Why are people spread unevenly?

When people first migrated across Africa, they settled where they could survive best. They needed land suitable for growing crops or raising animals, or access to the coast for fishing and trade. They needed water to drink, and rain for farming.

Much later, people moved to find work other than farming. Africa's cities grew, and are still growing – fast. Lagos, Cairo and Kinshasa are **megacities** – urban areas with more then ten million people.

▲ In the Makoko slum in Lagos. Here, along the waterfront, people live in shacks on stilts.

◀ Part of Lagos. It is Africa's biggest city. Over 2000 people a day move here from other parts of Nigeria.

Your turn

1 What does this term mean? (Glossary?)
   a population   b population density

2 Look at table **B**.
   a Which African country has the smallest population?
   b Which has the largest population?
   c About how many more people are there in Nigeria than in Ghana?
   d Compare the populations of Chad and Angola. (See page 16 for *compare*.)
   e The population of the UK is around 66.7 million. (That's 66 700 000 people.) Name the African countries which have a larger population than the UK.
   f The population of London is about 8 800 000. How many African countries have fewer people than London?

3 a From map **A**, identify one country with a high population density overall. (The map on page 106 gives names.)
   b Now identify one with a low population density.

4 Explain these facts. (Pages 110 and 112 will help.)
   a Africa is *sparsely populated* along the Tropic of Cancer. (Glossary?)
   b Most of Africa's coastal areas are quite *densely populated*.
   c People like to live around East Africa's great lakes.
   d There's a wiggly strip of high population density in Egypt.

5 Lagos, in Nigeria, is Africa's biggest city. It is growing very fast. People are moving in from rural areas. Look at this graph.

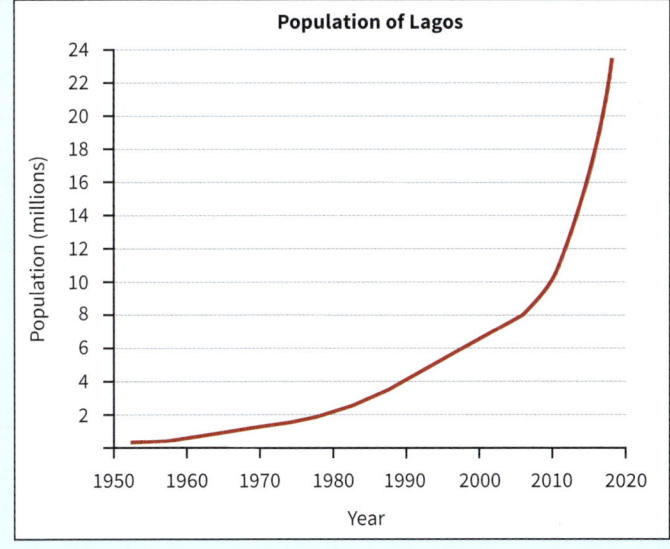

   a What was the population of Lagos in 1990? (About …)
   b What was its population in 2018?
   c Why might people want to move from rural areas to Lagos?
   d A fast rise in population can cause problems for a city. See if you can explain why.
   e Many new arrivals end up living in **slums**, like the one in photo **D**. What difficulties might they face?

109

6.6 What are Africa's main physical features?

 Now learn about Africa's key physical features – and where they are on the map.

Why ... ... does the Sahara get so little rain?

Did you know?
- The Bible and Koran tell how Moses led the Israelites out of Egypt, across the Red Sea.

Mountains, rivers, deserts, lakes ...

Study this map for a little while. Then try *Your turn*.

Key

land height above sea level in metres
- more than 2000 m
- 1000 – 2000 m
- 500 – 1000 m
- 200 – 500 m
- less than 200 m
- land below sea level
- ▲ highest peaks with heights in metres
- river
- lake
- country border

Did you know?
- The Kalahari Desert is really semi-desert.
- It has lots of vegetation after rain.

110

Your turn

The map on page 106 and the satellite image on page 99 will help you answer these questions.

1. The world's longest river is in Africa. Its name has four letters. It shows up clearly on the satellite image on page 99.
   a. Find it on the map, and give its name.
   b. It has two tributaries. Name each of them.
   c. Which sea does this river flow into?
   d. Describe what you notice where the river reaches the sea:
      i on the map ii on the satellite image
   e. Name two countries that this river flows through.
2. Now name five African rivers that flow into the Atlantic Ocean.
3. This photo shows the famous Victoria Falls.

   a. Which river are they on? Look on the map!
   b. They lie on the border between two countries. Name these.
4. Find Africa's biggest lake on the satellite image on page 99.
   a. Now identify it on the map, and state its name.
   b. Three countries share this lake. Name them.
   c. Who is the lake named after? Guess! (Think about Africa's history. Map **C** on page 107 may help too.)
5. a. Which part of Africa is the most mountainous? The north? The west? Be as clear as you can in your answer.
   b. Where in Africa are the Atlas Mountains?
   c. Name two other highland areas in Africa.
6. The photo below shows Africa's highest mountain.
   a. What is its name? Use the map to find out.
   b. Which country is it in?
   c. It's near the Equator. So why does it have glaciers on top?

7. Now name Africa's second and third highest mountains. (They are also marked on the map.)
8. This photo was taken in the world's largest hot desert.
   a. It is in Africa. What is its name? (6 letters!)

   b. Where in Africa is it?
   c. Is it mountainous? Flat? A mixture? Give your evidence.
   d. Not much vegetation grows there. Explain how you can tell this from the satellite image on page 99.
9. Many of the things you own and buy are made in China. They reach the UK by sea, in container ships.
   They usually take the route shown on the map below.

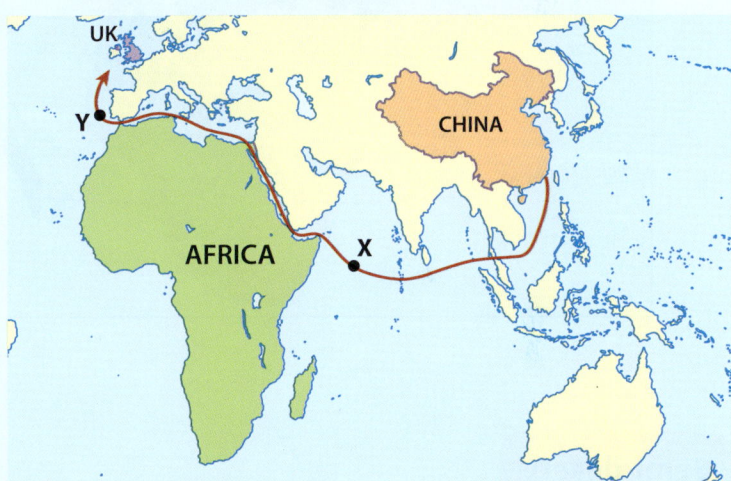

Using the maps on pages 110 and 106 to help you, describe the route a ship takes, to get from **X** to **Y**. Give the names of all the bodies of water it sails through. You could start like this:
The ship sails through the Gulf of Aden. Then it …

10. It's time for some work on your mental map of Africa. Practise drawing a rough sketch map of Africa. Be quick! Mark in and / or label these:

River Nile	at least two other rivers	two oceans
Red Sea	Mediterranean Sea	Suez Canal
Kilimanjaro	Lake Victoria	Sahara desert
Equator	Tropics	three mountainous regions

6.7 Africa's biomes

Here you will learn about Africa's four main biomes.

What's a biome?
A **biome** is a large region with its own distinct climate, plants, and animals. The climate dictates what the biome is like. That's because plants and animals adapt to suit the climate.

Africa's biomes
Africa has different climate zones – so it has different biomes. The panels tell you about the four main ones. Match colours.

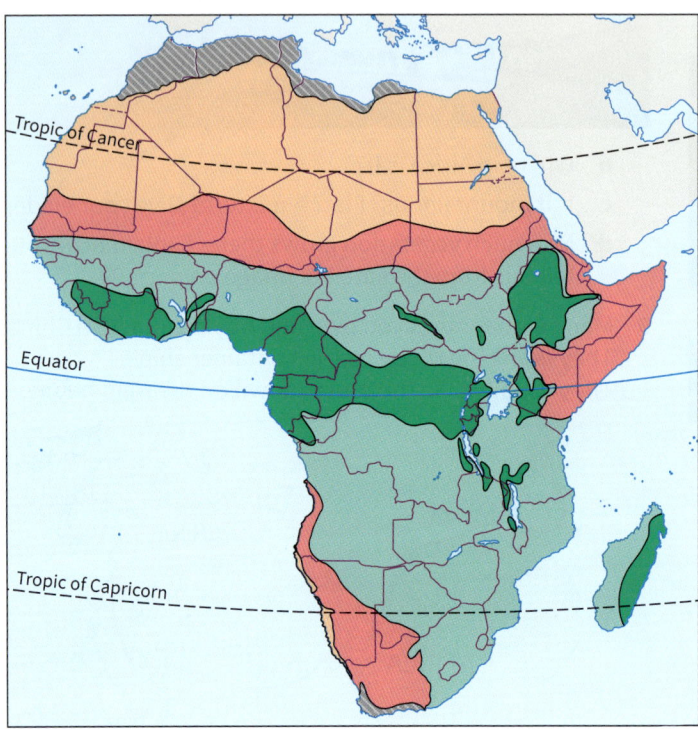

Hot desert

- Hot in the day – up to 50 °C. But very cold at night. (No clouds to keep the heat in.)
- Very little rain. Some places get almost none.
- Very strong winds.
- Plants must be able to find and store water, and protect themselves from heat and wind.
- So you'll find low spiny shrubs with long roots and thick stems, and wiry grasses.
- There are camels, gazelles, and ostriches too. All have adapted to survive for long periods without drinking.
- There are other animals like vipers, desert hedgehogs and sand rats, that hide from the sun's heat in burrows.

Semi-desert

- This biome lies between the desert and the savanna, and is always warm or hot. (But a bit cooler than the desert.)
- There's some rain over a few months of the year – but you cannot depend on it. The rest of the year is very dry.
- It has grass, and low shrubs, and scattered trees.
- You might see wild dogs, and lots of rodents.
- Life is hard here. People raise animals: cattle, goats, sheep, camels. Some grow crops such as maize, if they can water them. But rains often fail, so animals and crops may die.

Africa

Savanna

- The savanna is warm all year, with a rainy season (or sometimes two) and little rain the rest of the year.
- It is rolling grassland, with scattered trees (mostly acacia trees, as shown above).
- You may see lions, elephants, zebra, giraffes, and more. (Africa's game parks are in this biome.)
- Most people in the savanna raise animals, and grow crops where they can. In many places the soil is worn out and useless after years of overgrazing and growing crops. **Desertification** is a problem.

Rainforest

- Warm and wet all year - but less rain than in the Amazon rainforest.
- There are thousands of species of plants: from low shrubs and ferns to trees up to 45m tall.
- Animals include chimps and gorillas, many kinds of monkey, snakes, hippos, and hundreds of species of bird.
- Much of the African rainforest has been destroyed. People chop down trees for timber and firewood, and to clear land to grow crops.

Your turn

1 What is a *biome*?

2 Let's start with the hot desert biome.
   a It does not have much vegetation. Why not?
   b Describe two ways in which plants have adapted to the climate in this biome.
   c Explain how hedgehogs and rats can survive here.
   d The Sahara lies in this biome. Which countries share the Sahara? List them. (Page 106.)

3 a State two differences between the semi-desert biome and the hot desert biome.
   b The semi-desert region south of the Sahara is called **the Sahel**. Which countries lie (at least partly) in the Sahel?

4 You are on safari, visiting a game park in Africa.
   a i   Which biome are you in?
      ii Name five countries which lie in this biome.
   b Name three species of wild animals that you might see.

5 Which biome has the most lush vegetation? Explain why.

6 Now list the four biomes in the overall order in which they appear, when you travel north from the Equator.

7 Look at the satellite image on page 99. The deeper the green, the more vegetation there is. Compare the pattern with the map on page 112. Is there a match? Describe everything you notice.

8 Next, compare the maps on pages 108 and 112.
   a i   Overall, the hot desert biome has fewer than one person per square kilometre. Explain why.
      ii But some areas of this biome do have more people. Suggest a reason. (What do they need, to survive?)
   b Usually, few people live in rainforests, because of the dense vegetation. But many people are now living in Africa's rainforest biome. Explain what has happened to the vegetation, and why.
   c Describe the population density in the savanna biome.

9 Which of the four biomes would you like to spend:
   a most time in?   b least time in?
   Give your reasons.

6 Africa

How much have you learned about Africa? Let's see.

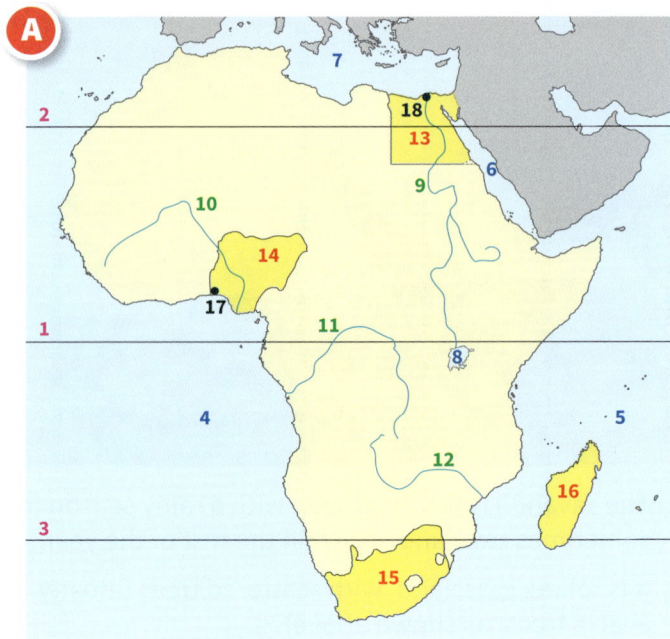

1 Map **A** shows Africa.
  a Name the line of latitude labelled: **i** 1 **ii** 2 **iii** 3
  b Identify the body of water labelled:
    i 4 **ii** 5 **iii** 6 **iv** 7 **v** 8
  c State the names of the rivers with these labels:
    i 9 **ii** 10 **iii** 11 **iv** 12
  d Identify the country labelled:
    i 13 **ii** 14 **iii** 15 **iv** 16
  e Which number on the map refers to:
    i Lagos, the largest city in Africa?
    ii Cairo, the second largest city in Africa?
  f Now look at the country labelled 13.
    i Over 90% of this country is almost empty of people. (See the map on page 108.) Explain why.
    ii Name three animals you are likely to find in the almost-empty area of this country.
    iii The population density is high along this country's big river. Suggest a reason.

2 African countries export materials to other countries.
  a Define *export*.
  b Map **B** shows the top export for each country in a recent year. (It is the one which brings in most money.)
    Name two countries which have this as their top export:
    i diamonds **ii** cotton **iii** oil **iv** gold
  c Of the top exports, identify:
    i five which are cash crops
    ii four which occur naturally in the ground
  d One country depends on tobacco as its main export.
    i Identify this country.
    ii Predict two consequences for its tobacco farmers, if everyone in the world stops smoking.
  e Although they export lots of resources, African countries often struggle to make money from them.
    i Suggest a reason. (Page 105?)
    ii Suggest one way that a gold exporting country could *add value* to its gold. (Page 105?)

3 Table **C** gives data for some countries.
  a Name the country or countries with:
    i over half **ii** over three-quarters **iii** less than 1%
    of the population living in extreme poverty
  b Which country is closest to eradicating extreme poverty?
  c *Some African countries are much better off than others.*
    Give evidence from the table to support this statement.

4 *If Africa had more factories, it would be better off.*
  To what extent is this statement true? Write at least 10 lines, giving reasons to support your answer.

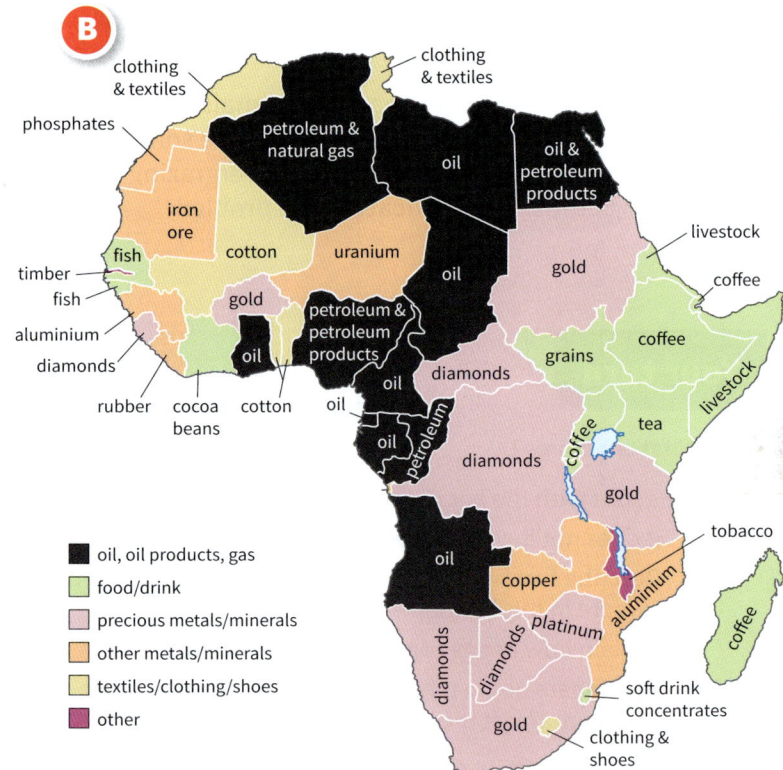

Country	% of the population living in extreme poverty
Algeria	0.5
Burkino Faso	43.7
Democratic Republic of Congo	77.1
Ghana	12.0
Malawi	71.4
Nigeria	53.5

7 Kenya

7.1   Hello Kenya!

This unit will give you an overview of Kenya, a country in East Africa.

> **Did you know?**
> • Kenya is one of the world's top destinations for safaris.

Where's Kenya?

▲ Kenya's flag. Black for the people, red for their struggle against the British, and green for Kenya's natural wealth. In front: a Maasai shield and spears.

Kenya is a country in East Africa. It lies across the Equator, and borders the Indian Ocean. It is shaped a bit like a pentagon. Its capital is Nairobi.

Look at the disputed area on map **B**. It is called the **Ilemi Triangle**. Kenya controls it, but South Sudan claims it too.

What's it like?
- Kenya has snowcapped mountains, volcanoes, lakes, beaches, forests, savanna, rich farmland, and a desert.
- It is usually warm or hot, and mostly dry.
- It has hundreds of species of birds and fish. And it has lions, leopards, elephants, giraffes, rhinos, zebras, wildebeest, crocodiles, and more.

▲ In a high-tech company in Nairobi.

Who lives there?
Kenya has a population of around 51 million people. (The figure is for 2019.)
- The number is rising fast. It is expected to reach 65 million by 2030.
- Half the people in Kenya are under 20 years old. So it's a **young population**. (Around a quarter of the people in the UK are under 20.)
- Kenya has over 40 African ethnic groups, each with its own customs and language. And non-Africans too: Arabs, Asians, and Europeans.
- The two official languages are English and Swahili.
- Around 85% of the population is Christian, and about 10% is Muslim.

▶ This man is from the Maasai ethnic group. The Maasai make up about 2% of Kenya's population. They live around Kenya's most famous wildlife reserve, the Maasai Mara. In the past they were famed as warriors.

Kenya

▲ We might look cuddly – but we'd eat you up.

▲ What about a camel ride beside the Indian Ocean?

What do people do?

- Over 60% of Kenyans live by farming. Shops in the UK sell some of the crops they grow – like tea and coffee, and vegetables, and flowers.
- The scenery and wildlife and friendly people attract hundreds of thousands of tourists every year. Tourists need food, and transport, and guides, and places to sleep. So thousands of Kenyans work in tourism.
- Some people work in manufacturing. The number of factories is growing.

Rich or poor?

Kenya has many wealthy people. But around 1 in 3 Kenyans lives in poverty, with barely enough to survive. Now the government is working hard to give everyone a better standard of living. Kenya's future looks bright!

▲ Picking tea leaves on a tea plantation. Some will end up in a cup near you.

Your turn

1. *Africa is a _____, and Kenya is a _____.*
   Copy and complete the sentence above, using words from this list: city   country   island   continent

2. Name the countries that share a border with Kenya.

3. This table compares Kenya and the UK:

	Kenya	UK
Area (sq km)	582 650	242 500
Population (millions)	51	67

   a i Which of the two countries is larger in area?
      ii About how many times larger is it?
         over twice as large   over 20 times larger
   b Compare their populations. (See page 16 for *compare*.)
   c Which country is more crowded? Explain your thinking.

4. What do the colours of the Kenyan flag represent?

5. Around what % of the population is aged under 20 in:
   a Kenya?   b the UK?

6. a Select one photo from this unit for a poster to attract tourists to Kenya. Explain your choice.
   b Now write a slogan about visiting Kenya, for the poster.

7. It's mental map time!
   a First, study map **A** on page 116 for a few minutes.
   b Now close the book.
      i On a page of your exercise book, sketch the outline of Africa. Do it roughly, in less than 2 minutes.
      ii On your sketch map, mark in and label:
         the Equator   Kenya (shade it in)   the Indian Ocean
   c Assess your map. How good is it?
   d Repeat **a** – **c**. See how much you can improve.

7.2 What are Kenya's main physical features?

Kenya has almost every physical feature you can think of! Find out more here.

Kenya's physical features

Mountains, glaciers, volcanoes, rivers, lakes, desert, beautiful beaches … Kenya has them all.

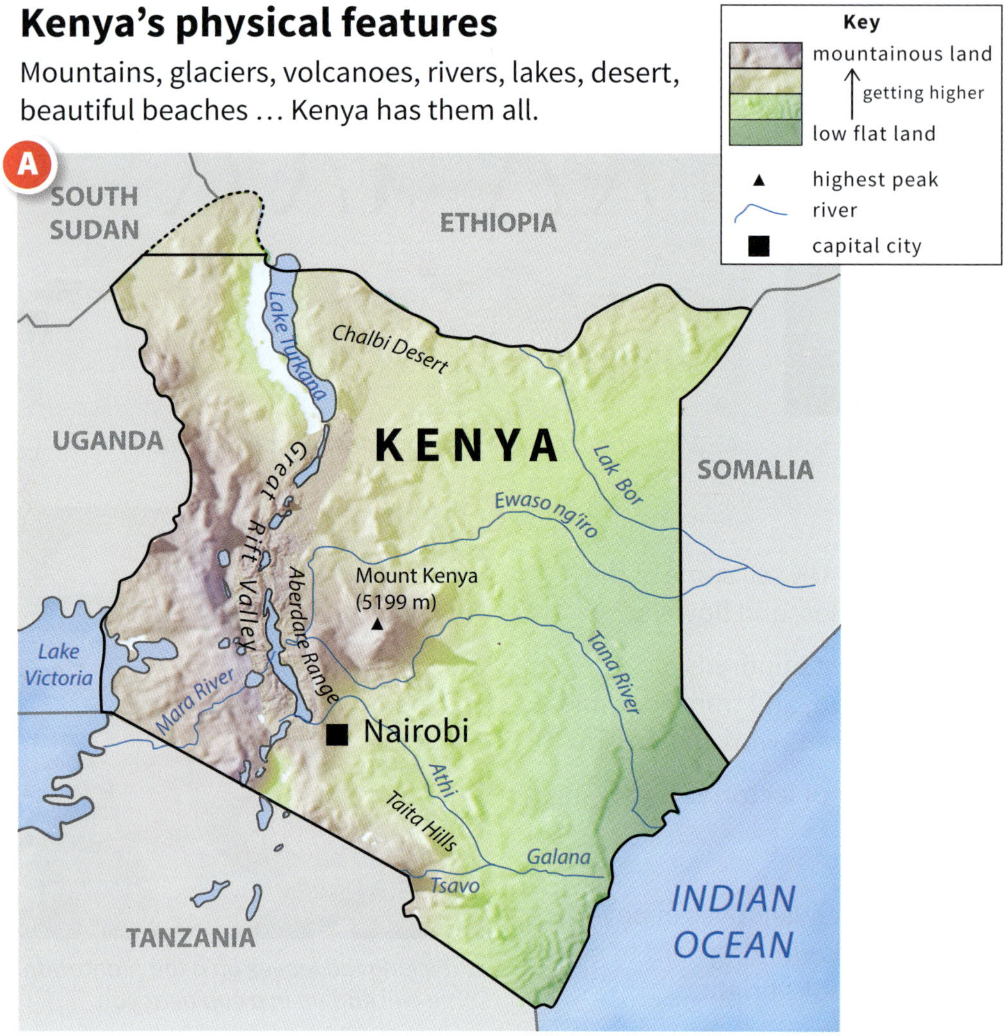

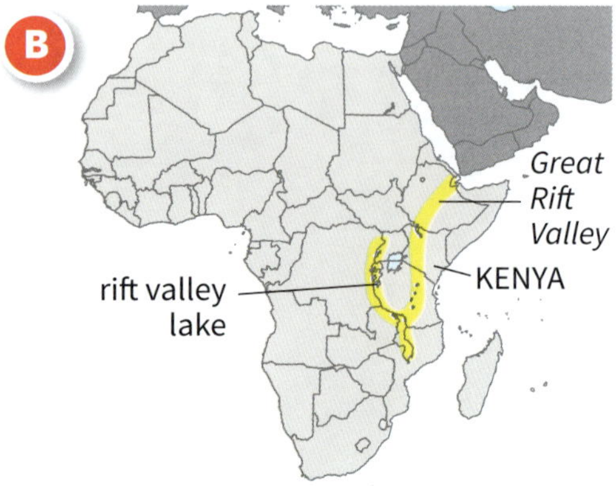

A [Map of Kenya showing South Sudan, Ethiopia, Uganda, Somalia, Tanzania, Lake Victoria, Lake Turkana, Chalbi Desert, Great Rift Valley, Aberdare Range, Mount Kenya (5199 m), Mara River, Ewaso ng'iro, Lak Bor, Tana River, Athi, Taita Hills, Galana, Tsavo, Nairobi, Indian Ocean]

Key
- mountainous land
- getting higher
- low flat land
- ▲ highest peak
- river
- ■ capital city

B The Great Rift Valley runs down through East Africa.

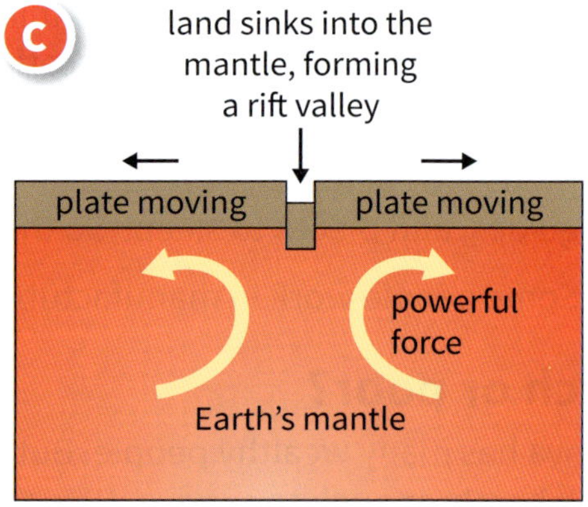

C How a rift valley forms.
- land sinks into the mantle, forming a rift valley
- plate moving ← → plate moving
- powerful force
- Earth's mantle

As you can see from map **A**, the eastern half of Kenya is quite low and flat. The high and mountainous lands lie in the western half. They are Kenya's **highlands**.

The Great Rift Valley

The Great Rift Valley is a huge trench that runs down through Kenya. Find it on map **A**. In fact it runs through several African countries. Look at **B**.

The rift valley formed because massive slabs of Earth, called **plates**, are being dragged apart. Look at diagram **C**. The rift is slowly getting wider, and the valley floor is still sinking.

Experts say that millions of years from now, all of Africa's Great Rift Valley will fill with ocean water. When that happens, a chunk of East Africa will become an island!

▼ Part of Kenya's Great Rift Valley. A string of volcanoes and lakes lies along the valley, linked to the rifting process.

Mountains and volcanoes

- Mount Kenya is Kenya's highest mountain. It is an **extinct** (dead) volcano, with glaciers.
- There are many volcanoes along the Great Rift Valley. (The forces that cause rifting also produce volcanoes.) Some are extinct, but most are **dormant** (sleeping), and could erupt one day.

Rivers and lakes

- Map **A** shows only Kenya's longest rivers. There are many more. Some small rivers dry up in the dry season.
- Look at Lake Turkana. It is the largest of eight **rift lakes** that lie in Kenya's Great Rift Valley. These lakes were formed by rainwater flooding into the valley.
- Lake Victoria lies west of the Great Rift Valley. It is the largest lake in Africa, by area. Kenya shares it with two other countries.

Dry lowland, and desert

Look at Kenya's low land, on map **A**. It is mainly flat plains, dry or very dry, with grass or low shrubs. The Chalbi area is so dry that it's a desert.

▲ Mount Kenya – Kenya's highest mountain. (Kenya is named after it.)

▲ The remains of an extinct volcano at the edge of Lake Turkana.

▲ In the Chalbi desert.

▲ The Tana is the longest river that's all within Kenya – 1000 km.

Your turn

1. a A large trench runs down through Kenya. Name it.
   b Draw a labelled sketch to show how this trench formed.
2. a What is a *physical feature*? (Glossary?)
   b Name two types of physical feature found within the Great Rift Valley.
3. Which countries share Lake Victoria? (Map on page 116?)
4. a Name Kenya's highest mountain.
   b This mountain is only 20 km from the Equator, where it's hot. It has something surprising on it. What? And why?
5. Look at the rivers on map **A**.
   a Identify the longest one that touches no borders.
   b Name the mountain range where this river rises.
6. Using map **A** to help you, describe where the high land is, in Kenya. Include all these terms: *west   south west   5100 m   Mount Kenya   Aberdare Range   Nairobi*
7. Kenya has a coast. Suggest two ways this may benefit Kenya.
8. *Kenya has almost every physical feature you can think of! Justify this statement.*

7.3 What's Kenya's climate like?

 Here you'll learn more about Kenya's climate, and climate zones.

Climate

Climate means what the weather in a place is usually like: how hot or cold it is, and how much rain it gets.

How hot is it in Kenya?

Kenya lies across the Equator, where the sun's heat is strongest. So most parts of Kenya are hot all year round.

But temperature also depends on **altitude**. The higher you go, the cooler it gets. So Kenya's highlands are cooler.

Map **A** shows how average temperature varies across Kenya. Compare this map with map **A** on page 118.

How much rain?

Map **B** shows how rainfall varies across Kenya.

But the rain does not fall evenly through the year. Most falls in **rainy seasons**. Much of Kenya has two rainy seasons, one in March – May and a lighter one in October – December.

> **Did you know?**
> • Rainy seasons are due to changes in wind direction.

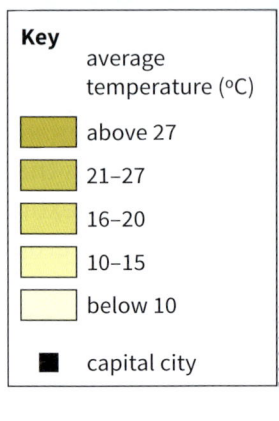

◄ Kenya's average temperature (°C).

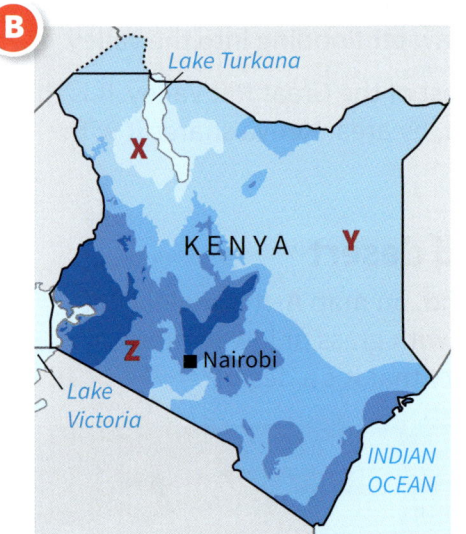

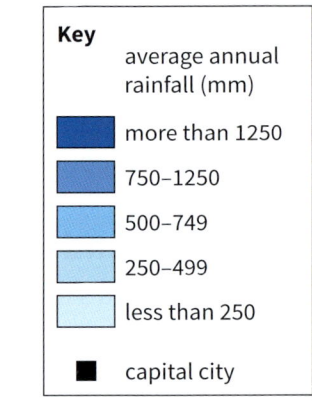

◄ Kenya's average annual rainfall (mm).

Kenya's climate zones

Combining temperature and rainfall data gives us **climate zones**. Look at map **C**.

① **Hot and very dry.** These are arid lands. (*Arid* means *very dry*.) Growing crops is very difficult. So people are **pastoralists** – they rear animals.

② **Hot and dry.** These are **semi-arid** lands. Here you'll find **savanna**: grassy plains with scattered trees. People rear animals, and grow crops only if they can water them.

③ **Cooler and moister.** With enough rain, and fertile soil, the highland areas are good for growing crops: tea, coffee, vegetables, maize, and so on.

④ **Hot and humid.** This is low flat land along the coast. It has swamps and mangrove forests. Coconut palms are grown, and many types of fruit.

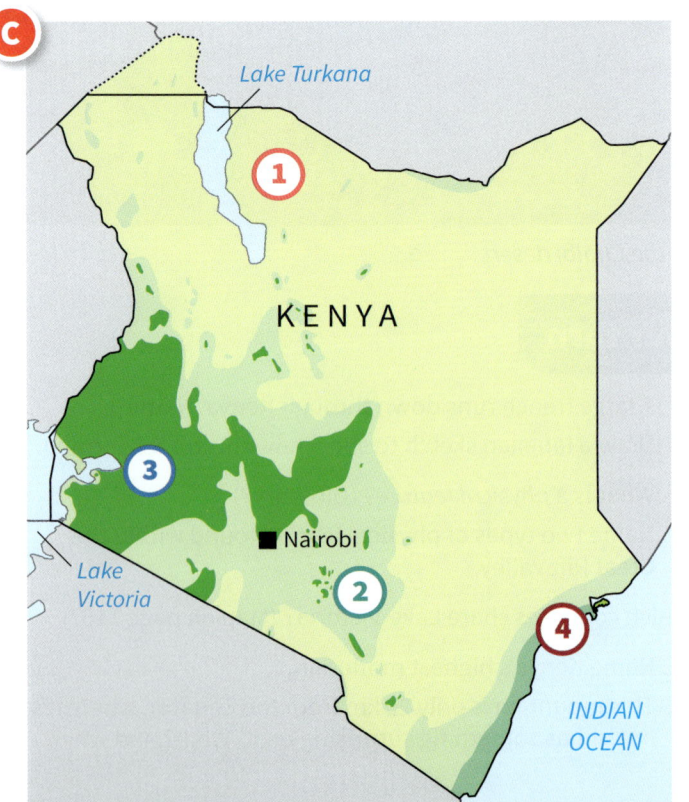

Kenya

▲ The savanna. Look closely. What's in that wavy loop?

▲ Kenya's highlands. Rain plus fertile soil means good farming.

Why Kenya suffers drought – and flooding

Around 80 % of Kenya is dry or very dry, as map **C** shows. So drought is a constant threat – but so is flooding! Here's why.

- The rains are unreliable. They may fail for several seasons in a row, causing drought. Vegetation and crops shrivel. Animals and people die.

- Sometimes the opposite happens: very heavy rain, even in arid areas. Sudden **flash floods** wash away animals, people, crops, and homes.

- Drought and floods are now occurring more often in Kenya, because of **global warming**. (As Earth gets warmer, rainfall patterns change too.)

- People are trying to prepare for this. For example, by finding ways to store rainwater, and growing crops that cope better with drought.

▲ A mangrove forest at Kenya's coast. Fish breed among the tree roots.

Your turn

1. Map **A** shows Kenya's average temperature. Look at the key. **P**, **Q** and **R** on the map represent places.
   a. In which place is the temperature in the range 10 – 15 °C?
   b. What is the temperature usually like at **R**?
   c. **P** and **Q** are the same distance from the Equator. Explain why it's cooler at **P** than at **Q**. (Map **A** on page 118 will help!)

2. Now look at map **B**.
   a. Which place gets least rain: **X**, **Y** or **Z**?
   b. *Overall, the northern half of Kenya is drier than the south.* Is the statement in italics true, or false?

3. Rain is not spread evenly through the year in Kenya. There are **dry seasons** with little rain. Suggest one disadvantage of this.

4. Explain why Kenya's highlands are the best region of Kenya for growing crops.

5. In fact less than 20% of Kenya's land has a suitable climate, and fertile soil, for growing crops. This is a disadvantage for Kenya. See if you can suggest why.

6. Kenya has to cope with two very different hazards, linked to rainfall.
   a. Name the two hazards.
   b. Both can lead to loss of life. Explain why.
   c. These hazards are likely to cause even more suffering in the future. Why?

7. Study photos **D** and **E**. Then write at least five sentences comparing the landscapes in these photos.

7.4 A short history of Kenya

Kenya was once a British colony. Find out more.

Up to independence

5000 years ago, the land we call Kenya was empty of humans (*Homo sapiens*) except perhaps for small groups of **hunter-gatherers**.

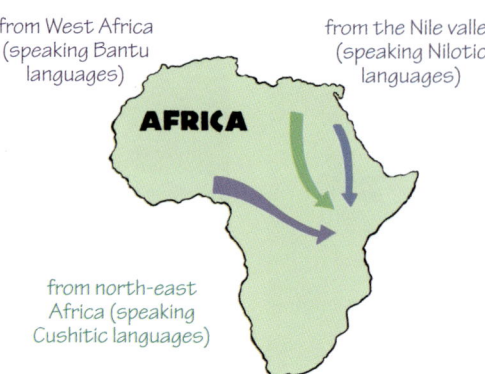

But slowly, more humans arrived – from three regions that had many ethnic groups. (That's why Kenya has so many ethnic groups today.)

Each group brought its own customs. The Cushitic and Nilotic groups reared animals. The Bantu groups grew crops – and knew how to extract iron.

Some Bantu settled at the coast. By 800 CE they were already trading with Arabs who sailed across the Indian Ocean from Oman and other countries.

In 1498 the Portuguese arrived. They fought with the Arabs for control of the coast and sea routes. The Arabs finally drove them out in 1730.

In the 19th century the British arrived. And in 1884 Britain was given rights over the area that's now Kenya, at the **Berlin Conference** (page 102).

The British pushed inland from the coast, taking control of the area. And in 1920, they made it a British **colony**. They named it **Kenya**.

Kenya's highlands have very fertile land. The British threw the Africans off this land, and offered it to white Europeans.

The Africans resented how they were treated. And in 1947 different ethnic groups came together and set up a political party to oppose the British.

Kenya

They had little success. So in 1952 the **Mau Mau** began a fierce armed revolt. They killed white settlers, and Africans who worked for the British.

In 1960, after eight years of fighting, the British Army finally defeated the Mau Mau. But it was clear that Britain could no longer control Kenya.

So on 12 December 1963 Kenya gained independence. The first president was a Kikuyu man called Jomo Kenyatta. Many Europeans soon left.

Kenya today

Now Kenya rules itself.

- It still has strong links with the UK. It's part of the **Commonwealth**.
- It has strong links with other countries too. China runs big projects in Kenya, building new roads and railways, and a new port. It sends Chinese experts and workmen. It lends Kenya money to pay for the projects.
- Kenya is **developing** quite fast. (*Developing* means changing in ways that improve people's lives.) It is getting better off year by year.
- But many Kenyans are still living in deep poverty.
- When Kenya's borders were drawn, different ethnic groups were forced together. They did not get along, and there is still conflict between them. Sometimes this leads to violence.
- The fertile farmland is still not shared fairly. Some is still owned by white settlers. Large tracts are owned by the families of Kenyan politicians who bought land from white settlers.

But overall, Kenya has a bright future. You can find out more about how it's doing, and the challenges it faces, in Unit 7.10.

Did you know?
- The Maasai are a Nilotic ethnic group. The Kikuyi are Bantu.
- Both were moved off their traditional lands by the British.

Did you know?
- Since 2010, foreigners can't buy farmland in Kenya.
- If they already have some, they must give it up after 99 years.

Your turn

1. The African people of Kenya originated from three regions of Africa. Name these.
2. a. A Bantu language called **Swahili** is one of Kenya's official languages. It contains some Arabic words. Explain why.
   b. The other official language is English. Suggest a reason.
3. a. Portuguese ships did not arrive at East Africa until centuries after the Arab traders. Suggest a reason. (The world map on pages 140 – 141 may help.)
   b. Why were the Portuguese so interested in East Africa?
4. a. When was Britain given rights to the area that's now Kenya?
   b. Give a reason why some of Kenya's borders are straight.
   c. For how long was Kenya a British colony?
   d. Who were the Mau Mau?
   e. Why did the Mau Mau fight the British?
5. The second '*Did you know*' above shows Kenyan laws made in 2010. Suggest a reason why these laws were made.
6. Now write a summary of Kenya's history. You can present it as bullet points, or a flowchart. Give dates. Keep it short!

7.5 Kenya's population

Kenya has a young population, and it's growing fast. Find out more here.

Population

Population means the number of people living in a place.

As you saw on page 122, Kenya was almost empty 5000 years ago. But slowly, over time, people arrived from three other African regions.

- Today Kenya has a population of around 51 million (the figure for 2019).
- There are over 40 African ethnic groups, each with its own customs and language. (Some groups are very small in number.) There are non-Africans too: Arabs, Asians, and Europeans.

▲ A Kikuyu man dressed in traditional clothing, for a celebration. The Kikuyu are Kenya's largest ethnic group. They make up 22% of the population.

Where does everyone live?

Population density means the number of people living per square km in a place.

Map **B** shows the population density for Kenya. The darker the shade on the map, the more people live there. The map shows only the biggest cities and towns. (Kenya has many other towns too, and thousands of villages.)

As you can see, some areas of Kenya are more **densely populated** than others. Some areas have very low population density. Why?

Map **A**, for rainfall, might give clues. Compare **A** and **B**!

Did you know?
- About 31% of Kenya's population is urban – living in cities and towns.

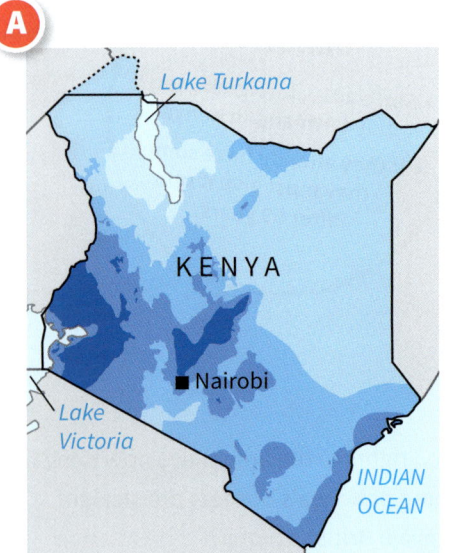

▲ Average annual rainfall in Kenya. The darker the shade the more rain there is. (See the key on page 120.)

Did you know?
- About 70% of Kenya's population is from Bantu ethnic groups.

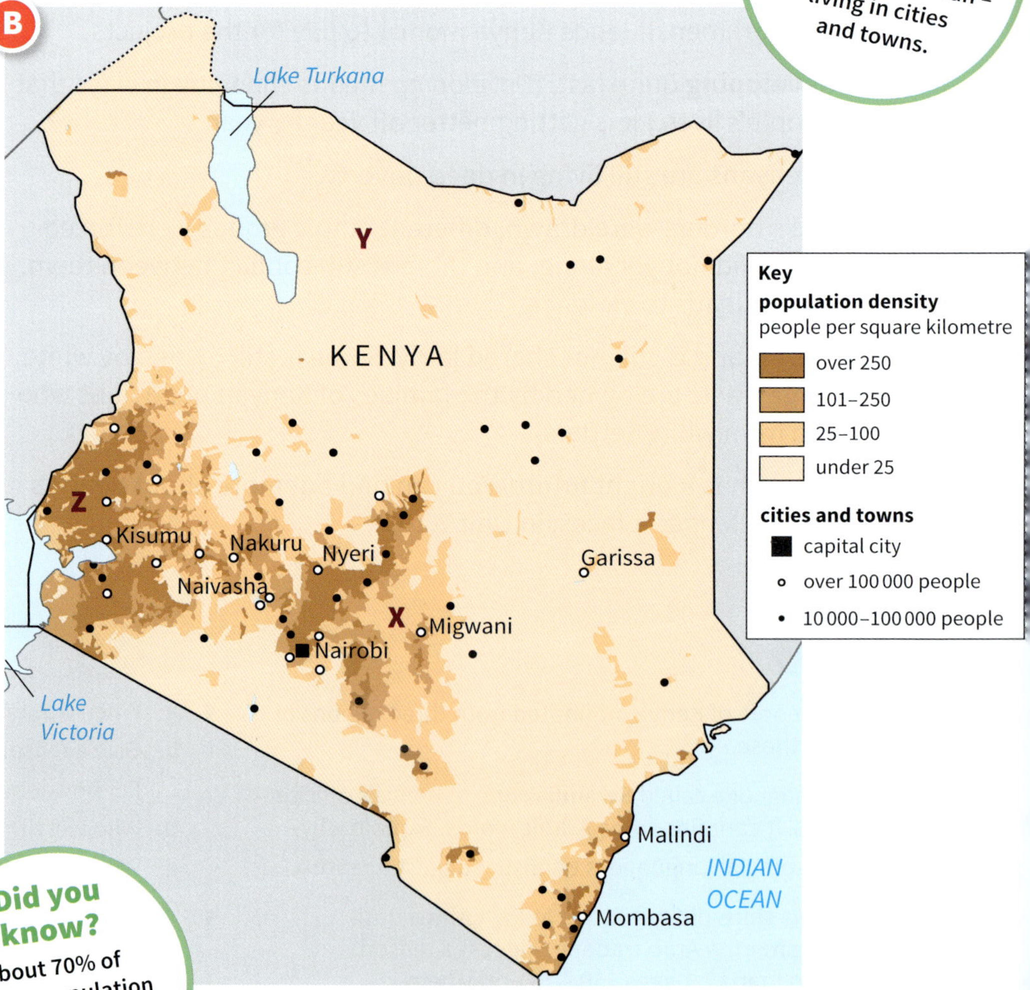

▲ Population density in Kenya. The darker the shade the more people there are.

Key
population density
people per square kilometre
- over 250
- 101–250
- 25–100
- under 25

cities and towns
- ■ capital city
- ○ over 100 000 people
- • 10 000–100 000 people

Kenya

The population is rising fast

In the year 2000, the population of Kenya was about 31 million. By 2050, it is predicted to be about 95 million. 64 million more people in just 50 years!

Two reasons for this fast rise are:

- better **healthcare**. There are more doctors and clinics, so people are living longer. Fewer new babies die.
- quite a high **fertility rate**. Each woman has 4 children, on average. (Some have more, some fewer.)

Lots of young people

Graph **D** is a special type of graph called a **population pyramid**. It shows the % of Kenya's population in each age group in a given year.

Look at the labels on the axes, and the key.

Now look at the lowest bar. It's for the age group 0 – 14. The girls in this group made up 20.2 % of the population. The boys made up 20.5%.

So 40.7 % of Kenya's population was aged under 15 that year. (20.2 + 20.5 = 40.7)

In the UK, 17.9 % of the population was under 15 that year. Compared with the UK, and many other countries, Kenya has a **young population**.

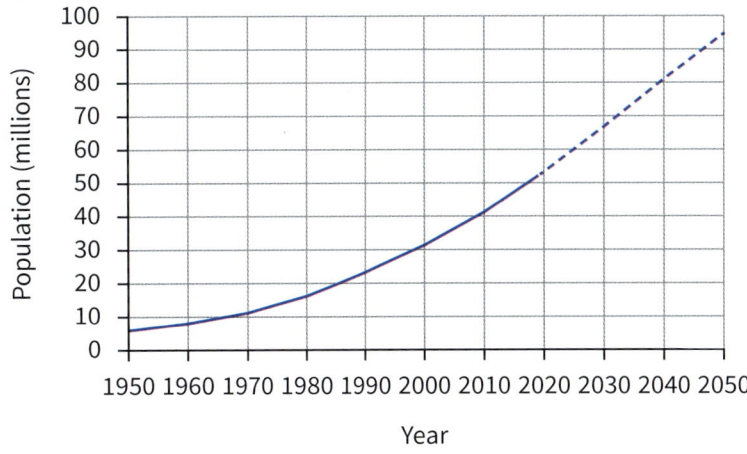

C Population growth in Kenya

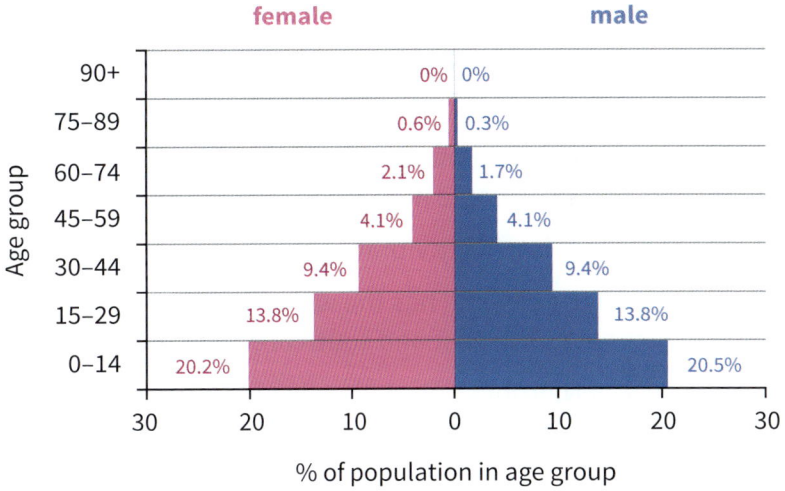

D A population pyramid for Kenya, 2018

Your turn

1. Define: **a** population density **b** densely populated

2. Look at places **X**, **Y**, and **Z** on map **B**.
   - **a** Which place has at least 250 people per square kilometre?
   - **b** Which has a higher population density, **X** or **Y**?
   - **c** State the range for the population density at **X**. Don't forget to give the units, *people per sq km*.

3. Compare the patterns in maps **A** and **B**. Then write a paragraph to explain why the population density at **Y** is low. (*Hint*: we need water for drinking and crops and …)

4. Kenya's biggest settlements are shown on map **B**.
   - **a** Name Kenya's capital city.
   - **b** Two of the named settlements are ports. Identify them.
   - **c** Identify the named city that is:
     - **i** furthest west **ii** almost directly north of Nairobi

5. Look at graph **C**.
   - **a** What was Kenya's population in 1950? About … ?
   - **b** Around which year did the population reach:
     - **i** 20 million? **ii** 40 million?
   - **c** What has been happening to the population since 1950?
   - **d** Explain why part of the graph line is dashed.

6. As the population rises more schools are needed, and doctors, and so on. List ten other key things that Kenya will need more of, in the years to come.

7. Look at the population pyramid in **D**, for 2018. What % of Kenya's population that year was:
   - **a** men aged 30 – 44? **b** women aged 30 – 44?
   - **c** people aged 30 – 44?

8. The % of the population aged under 15 in Kenya is still rising. Suggest two problems this may cause for Kenya.

7.6 What's Nairobi like?

Nairobi is less than half the size of London, but growing fast. It is a city of great contrasts. Find out more here.

How Nairobi began

200 years ago, Nairobi was open land by a river, where the Maasai brought cattle to drink. They called it *Enkare Nyorobi*, or *place of cool waters*.

Roll on to 1889. The British are building a railway from Mombasa to Lake Victoria. They have brought in Indian labourers. When the line reaches Enkare Nyorobi they set up a railway depot, and call the place **Nairobi**.

A settlement grows around the depot ... and grows, and grows. In 1907, the British decide to make it their capital.

Look at the plan of Nairobi below, for 1927. The British live on the pleasant higher cooler land west of the railway. The Indians and Africans have their own areas in the east, where it's lower, hotter, and marshy.

Nairobi today

In 1963, when Kenya gained independence, Nairobi had 350 000 people. Today it has over 3.5 million – more than ten times as many.

Some of this rise is due to **natural increase** – women having babies. But most is because people **migrate** to the city from rural areas, to find work.

- Today, Nairobi is a lively vibrant city, and truly multi-ethnic.
- The west of the city still has the wealthier areas, where well-off people of all races live. The east and south are lower-income areas.
- Over 60% of Nairobi's population live in **slums**. Their homes are shacks with no running water or toilets, and often no electricity. People queue for public toilets and water taps.
- The fast rise in population has put a big strain on the city's **infrastructure**. There are problems with traffic, and electricity and water supplies, and sewage and rubbish disposal.

The government has a big plan to solve Nairobi's problems by 2030.

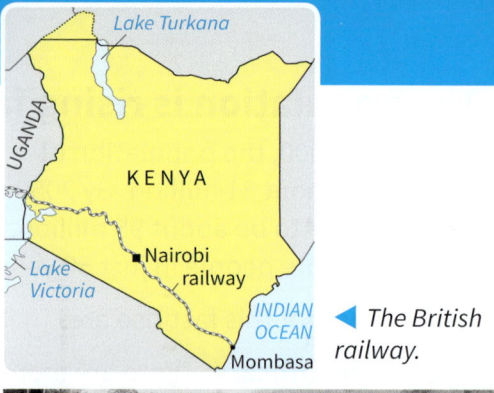
◀ The British railway.

▲ Nearly 37 000 Indians were brought in to build the railway. Most went home in 1901, when it was complete – but some stayed on.

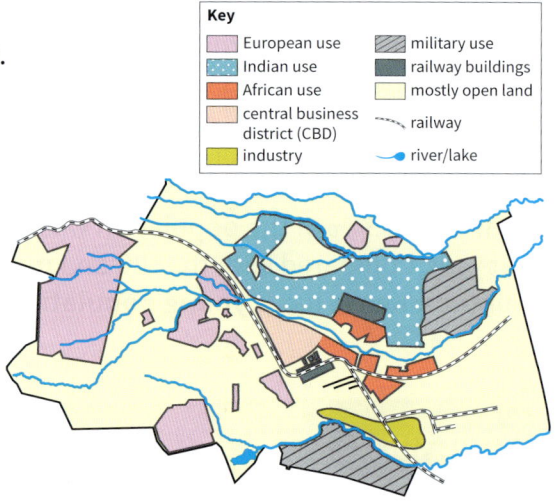
▲ A plan of Nairobi, 1927. Europeans, Indians and Africans have separate areas. (In 1927 the city had far more Indians than Africans.)

Your turn

1. Nairobi owes its birth to a railway. Explain why.
2. Look at the plan for Nairobi above, for 1927.
   a. Explain why the British chose areas to the west of the railway for themselves.
   b. Nairobi had thousands of Indians in 1927. Why?
3. The table below right gives population data for Nairobi.
   a. Draw a bar graph for this data. Use graph paper if you can.
   b. Now write a paragraph describing the population growth.
   c. State the main reason for this growth.
4. Look at the photos on page 127. Pick out:
   a. three photos that suggest a well-off city
   b. three that show poverty
   c. two that suggest problems with *infrastructure* (Glossary?)
   d. the photo that surprised you most – and explain why
5. You have rented a shack in Kibera. Write to your mum in Mombasa telling her what your shack, and Kibera, are like.

The population of Nairobi					
Year	1970	1980	1990	2000	2010
Population (millions)	0.5	0.9	1.4	2.2	3.2

Kenya

Nairobi – a city of contrasts

① Modern housing in the west of Nairobi. You will see lots of luxurious homes.

② The **central business district** or **CBD**. As in most big cities, it has tall office buildings.

③ A pleasant stroll on a sunny day. Lots of cafés and shops to visit.

④ A view from the Nairobi National Park. Wild animals – only 7 km from the city centre.

⑤ **Traffic congestion** is a big problem. The roads can't cope with all the traffic.

⑥ There are many small farms in the city. This man will sell what he grows.

⑦ Nairobi has over 60 slums. Above is Kibera, the biggest, with a quarter of a million people.

⑧ Many thousands of homes have no water on tap. This boy fetches it from an outdoor tap in a slum.

⑨ A shack in a slum. The shacks are built without permission. Most are rented out.

7.7 What does everyone do?

 People everywhere try to make a living, and feed their families. So what do people in Kenya do? Find out here.

Making a living in Kenya
Over 60% of Kenyans live by **farming**. (Many others farm part-time, alongside other jobs.) Some people are in **manufacturing** – making things in factories and workshops. The rest provide **services**. Bus drivers are an example.

1 Farming
Farming is the backbone of Kenya's economy.
- Many of Kenya's farmers are **subsistence farmers**. They grow crops just to feed themselves and their families.
- Some farmers are **pastoralists**: they rear animals – cattle, goats, camels, sheep.
- More and more of Kenya's farmers are growing **cash crops** – crops to sell. Like tea, coffee, vegetables, fruit. Cash crops are often exported.

The type of farming people do is linked to climate, as map **C** on page 120 shows.

More about the pastoralists
In Kenya's dry and very dry areas, most farmers are pastoralists. They feed their animals on free grass and other vegetation – so they must follow the rains. (When the rainy season begins, the vegetation grows really fast.)
- Some pastoralists are **nomads**. They are always on the move with their animals, carrying their belongings by camel.
- The rest are **semi-nomadic**. They travel with their animals to grazing, but have a homestead to return to. It may be close to a **borehole** or a stream. So they'll grow crops too, if they can water them.
- The pastoralists live on the milk and meat from their animals. They sell some when they need money for things like maize, and medicine.
- Life is getting harder for the pastoralists.
  - They need open land for grazing. But more and more land is being taken over, for example for game parks and windfarms.
  - Drought is the enemy – and climate change is bringing more drought.

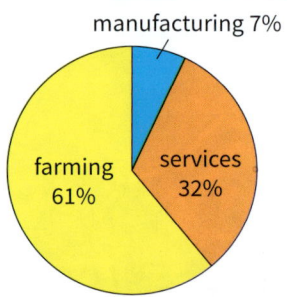

▲ How people in Kenya earn a living.

▲ Subsistence farming.

▲ Over 600 sq km of traditional grazing land were taken over for this big windfarm, near Lake Turkana.

◂ In Kenya's dry areas the need for grazing and water rules the pastoralist's life. Pastoralists include the Maasai in southern Kenya and the Turkana in the north.

▲ A plantation of pineapples, owned by a foreign company. It's near Nairobi. You'll find their tinned pineapples in a shop near you!

▲ A jeep from Kenya's first home-grown car manufacturer, Mobius Motors. That's the Great Rift Valley!

2 Manufacturing

Usually, in the early stages of a country's development, most people live by farming. Then as the country develops, factories are set up.

Kenya already has factories that make things like shoes, clothing, plastics, furniture, and cars. Some factories are owned by foreign companies.

Factories provide jobs. The goods can be sold to local people, and exported too. So the Kenyan government wants lots more factories!

3 Providing services

People who provide services include teachers, doctors, nurses, policemen, taxi drivers, market traders, and hair dressers.

Some people provide **high-tech services**. They do research, and develop tools, and provide information for other workers. For example some develop breeds of plants that need less water, for farmers.

Some high-tech services depend on mobile phones. 95% of Kenya's population lives in areas with a mobile signal. Most don't have bank accounts – and many don't even have electricity. But they can use mobiles to order and pay for goods, and get loans. They can get warnings about drought and floods. They can charge their mobiles using solar power.

▲ A service for sending money by mobile, developed in Kenya. You pay cash at the booth, then text a code to another person, who collects the money at another booth.

Your turn

1 These are from the terms in bold on page 128, about farming. Unjumble them, and then define them! (Glossary?)
   sach procs   sattoparsil   bussisceent merrafs   heelboor

2 Using map **C** on page 120 to help you, explain why:
   a most cash crops are grown in Kenya's highlands
   b most farmers in northern Kenya are pastoralists

3 Pastoralism in the dry areas is tough, and is becoming even more difficult. Give two reasons to explain why.

4 Which photo in this unit shows a cash crop?

5 a Look at the pie chart on page 128. Which type of work do:
       i most Kenyans do?   ii fewest Kenyans do?
   b Give one example of a job in manufacturing.
   c Give three examples of jobs in the service sector.

6 You are Kenyan. You are setting up a factory in Mombasa to make fridges. Write a blog for your website to tell everyone how the factory will help Kenya. At least 50 words.

7 Far more Kenyans have mobile phones than have electricity. Outline two ways in which mobiles improve their lives.

129

7.8 How Kenya earns money from flowers

 Kenya earns a lot by exporting cut flowers. Find out more here.

In a supermarket near you …

Most of Kenya's flower farms are around Lake Naivasha, in the Great Rift Valley. This is how the flowers reach a supermarket near you:

The flowers are cut. Some are made into bouquets, with supermarket labels on. They are put into boxes.

The boxes are taken to the airport in a cooled truck, and loaded into a cargo plane, bound for the UK.

Within 48 hours, the flowers are in pails in the supermarket, ready for shoppers to buy.

It's not just the UK. Flowers are flown as far as Australia and Japan. Many go to the Netherlands, where they are auctioned for sale across Europe.

A successful business

Kenya earns £ millions a year from cut flowers. Its flower farms employ over 90 000 people. They and their families depend on the flower business.

Why is it so successful around Lake Naivasha? Here are some factors …

 Locational factors
- sunshine all year, so flowers grow all year round
- not too hot, since the area is 1900 m above sea level
- fresh water from Lake Naivasha to water the flowers
- not far from Nairobi airport (1.5 hours by truck)

 Human factors
- easy to find workers (people move here to find jobs)
- people work hard
- wages are low here, so flowers can be grown for less than in other countries, and sold more cheaply

But flower farming is not all roses. Read the next page. Then do *Your turn*.

Your turn

1 Look at the factors listed above. They help to make flower farming around Lake Naivasha a success. Choose the three you think are the most important, and explain your choice.

2 The flower farms affect each of the groups in **a – e** below. For each, decide whether the overall impact is positive (helpful) or negative (harmful), and give at least one reason.
  a flower farm workers like Miriam
  b the local Maasai herders like Joseph
  c the local fishermen like Silas, who fish legally
  d shoppers in the UK
  e the Kenyan government, which taxes the flower farms

3 Now select one negative impact of flower farming around Lake Naivasha, and suggest a way to improve the situation.

4 *Although flower farming has some negative impacts, Kenya should keep on with it.*
  Do you agree? Decide, and then justify your decision.

Kenya

It's not all roses!

Miriam, a flower packer

We came here to find work. We have three children. We live in a little house in Karagita, near the lake. Lots of the flower farm workers live there. I earn £240 a month.

It's hard work. But at least the company sends free buses to pick us up and bring us home. And it runs a school. That's the best thing. I want education for my children so they can get better jobs.

Many flower farms are not so good. Some pay less than £150 a month. You can't live on that. And some workers get ill from the stuff they spray on the flowers.

Joseph, a Maasai herder

The flower farms are not good for us herders.

This used to be our grazing land. Our cattle need grass to eat, and lake water to drink. But more and more land is being taken for the flower farms, and workers' houses. Now it's even hard to find a passage to the lake!

So I might have to give up herding one day. We've always been herders. My family would be ashamed.

What other work can I get? Most of the farms are run by Kikuyu. They think we Maasai are good for nothing but cattle herding.

Silas, a fisherman

I've always fished. I have a boat and a licence. But now it's harder to make a living from fishing.

First, the water level in the lake has dropped, because the flower farms pump out water for the flowers.

Next, the lake is polluted. With fertiliser from the flower farms, and sewage dumped by the people in Kaitanga. These harm fish and make water weeds grow.

And third, illegal fishing. People come from all over to find work on the flower farms. If they fail, they fish. They use little nets and catch the young fish, which don't get a chance to grow. So … fewer fish for me.

7.9 On safari!

Kenya is rich in wildlife, which tourists pay to see. But tourism brings conflicts. Find out more here.

Kenya's wildlife

You want to see wildlife? Go on safari in Kenya! See lions, leopards, hippos, rhinos, elephants, giraffes, monkeys, wildebeest, crocodiles, and more.

Kenya has 65 **national parks** and **reserves**, where wildlife is protected. Some are lakes. The map shows the main ones. You can stay in lodges or camps, and travel with guides.

Wildlife and conflicts

Wildlife tourism is not all sunshine. Look at the conflicts around the Maasai Mara game reserve, for example. It's in the area where the Maasai live.

▲ Kenya's top national parks and reserves.

A The government
First, the government needs money. It can earn some from tourism. And tourism creates jobs.

Speech bubbles: We need money to run Kenya. For schools, hospitals, police … / Tourists bring in money. / Tourists like wildlife. / We need more game reserves, to protect the wildlife. / We'll charge a tax for every tourist. / Let's move the local people out of the way.

B The tour operators
They want to take as many tourists on safari as they can! The more tourists, the more profit they make.

Speech bubbles: I think I see a leopard. / Let's take a selfie. / But where are the lions? / Can't we get closer?

C The local people
The Maasai were cleared off land they had always used for grazing, to make way for the game reserve.

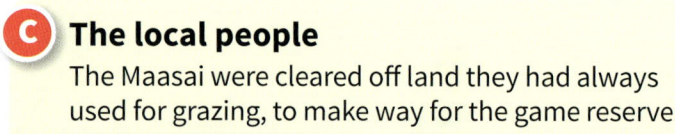

Speech bubbles: We're banned from grazing our cattle in the Maasai Mara. / They think wild animals are more important than us! / Sometimes the wild animals run out and kill our cattle … / … and if we shoot them we're fined. / The lodges make lots of money. We get none.

D The animals
They are under stress from tourist jeeps, and tourist lodges … and poachers.

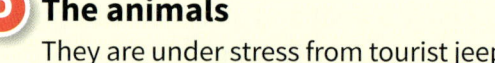

Speech bubbles: Jeeps ruin the grass. / The Maasai resent us. / Jeeps follow us everywhere. No peace. / And what about poachers? There used to be far more of us. / More lodges for tourists means less space for us to roam.

It's not just wildlife

Wildlife is not the only tourist attraction in Kenya.

There are white sandy beaches, and snorkelling. Rivers for whitewater rafting. Mount Kenya to climb. Tourists often combine a wildlife safari with other activities, for a very exciting holiday.

In fact, Kenya depends on tourism. It is one of Kenya's top earners. And tourists love Kenya.

A large % of the tourists are from the UK.

Tourism: good or bad?

So is tourism good, or bad? It has positive and negative aspects:

▲ *Trekking on Mount Kenya. Most people take two days to reach the summit. You can stay in overnight camps.*

✓ positive aspects	✗ negative aspects
brings money into the country	the local people may see very little of the money; it may even leave the country if lodges and camps and hotels are owned by foreigners
provides jobs for local people	jobs are often the poorly paid ones (like waiters and cleaners)
tourists enjoy the attractions	the local people may resent tourists; and the tourists may harm the attractions they came to see (for example wildlife)

How can tourism be improved?

There are several ways to improve tourism. For example:

- involve local people in decisions (such as whether to create a new reserve)
- share profits fairly with local people (especially if they were moved off land)
- give local people the better-paid jobs too, with training
- control the number of tourists in a place, so that it does not get damaged.

Then tourism becomes more **sustainable**. That means it can be carried on into the future to everyone's benefit, without doing harm.

Did you know?
- Poachers kill elephants to get their ivory tusks.
- Ivory is used for jewellery and ornaments.

Your turn

1. Define these terms: **a** safari **b** tourist **c** game reserve

2. All these groups are affected by tourism in the Maasai Mara:
   the local Maasai   the wild animals   the government   the tour operators   the tourists
   a Which three groups do you think benefit most?
   b Which two groups do you think benefit least?
   In each case justify your answer.

3. Imagine you are one of the tourists in **B**. Write a review of your safari for TripAdvisor. Write at least three sentences. Then add a rating for it, from one star (awful) to five stars (great).

4. You are in charge of tourism in Kenya. You want to make it more *sustainable*. (Glossary?)
   a Describe what action you could take to prevent scenes like the one in **B** on page 132.
   b Now outline two things you'll do to improve life for the Maasai who live beside the Maasai Mara.

5. Think about this girl's opinion.
   a Do you agree with her? Decide.
   b Now justify your decision. Write at least five sentences.

It's better for the planet if we watch wildlife films at home, instead of going on safari.

7.10 So how is Kenya doing?

Let's step back and see how Kenya is doing, and how it plans to do better by 2030.

Rich or poor?

If you list Africa's 54 countries in order, from richest to poorest, Kenya comes about halfway. But compared with the world's rich countries, it is poor. Look at this table for a recent year:

Data for …	Kenya	Malawi	Nigeria	Botswana	UK	USA
Population (millions)	51	19	197	56	67	327
GNI per person (PPP)	$2960	$1060	$5230	$11 920	$39 120	$54 940

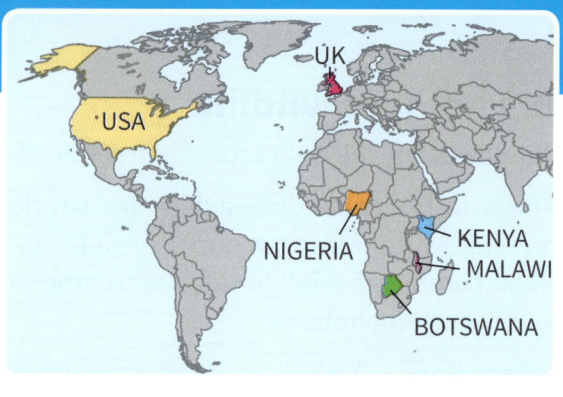

▲ The six countries in the table.

The bottom row indicates how well off people are.

- **GNI** stands for **gross national income**. This is a country's total income for a year. It is always given in dollars.
- GNI divided by the population gives you **GNI per person**.
- Things cost less in some countries than others. So GNI per person is now adjusted, to allow for this. The result is **GNI per person (PPP)**. It lets you compare countries fairly. (*PPP* stands for *purchasing power parity*.)
- A high value for GNI per person (PPP) suggests a high standard of living. People have a higher standard of living in Botswana than in Kenya.
- But remember, it is an *average* value. In every country – including the UK – some people are much richer than others. So there is inequality both between countries and *within* countries.

all the money a country's people and businesses earn, including from abroad

Inequality in Kenya

There are many wealthy people in Kenya, including several thousand super-rich. But look at these statistics for 2018:

Poverty in Kenya

- 44% of Kenyans have no electricity in their homes.
- 28% have no access to piped water, at home or in street pipes. They fetch unsafe water from wells or ponds or rivers.
- 41% have no access to a proper toilet or latrine.
- About 1 in 3 are below the poverty line. They live on less than £1.50 a day for food, clothing, medicine, transport, … everything.
- Poverty is greatest in the north and north east of Kenya.

So there are millions of very poor Kenyans. In fact Kenya is one of the most unequal countries in Africa.

▲ Some get their drinking water this way …

▲ … and some get it this way.

Kenya

Why is Kenya not better off?

Kenya has sunshine all year. A coast, for sea trade and fishing. Some very fertile land. Wildlife. Stunning scenery. A lively population. It has also received $ billions in **aid** from other countries.

So why is Kenya still quite poor? Here are reasons Kenyans give:

Did you know?
- Britain gives £ billions in foreign aid each year.
- Some goes to Kenya.

- Colonisation. The British took our best land for over 40 years. And did nothing much for us.
- Ethnic conflict. They put different ethnic groups together, to make Kenya …
- … but we don't all get along. Things often get violent.
- Not enough jobs. Lots of people have nothing to do.
- A rising population. More people to look after every year.
- Not enough rain. Most of Kenya is dry. The rains often fail. People lose everything.
- Healthcare still poor. People in poor health can't work.
- Not enough teachers. Many children leave primary school hardly able to read or write.
- Corruption. Aid money to help us goes into someone's pocket instead.
- Not enough factories. We should make more things for export.
- Lack of electricity, piped water, toilets. We'd work far better if we had decent homes.

But things are changing …

In fact Kenya is **developing** quite quickly, and the future is bright.

- The government has promised a better standard of living for everyone by 2030! Better healthcare, more jobs, more factories, new homes …
- Oil was discovered near Lake Turkana in 2012. The government can sell it to other countries, and use the money to help Kenya develop.
- China is playing a big part. It has lent Kenya money, and provided the expertise, for building new roads and a new railway. And a huge new port at Lamu on the coast.

Did you know?
- A new railway has been built from Mombasa to Nairobi to replace the British one.
- It was built by the Chinese and opened in 2017.

Your turn

1 Define these terms. (Glossary?)
   a standard of living
   b inequality

2 A high GNI per person (PPP) indicates that people are well off, on average. Look at the table on page 134.
   a In which of the six countries are people best off?
   b In which country are people poorest?
   c Is everyone in Malawi very poor? Explain.
   d People in the UK are about ____ times better off than people in Kenya, on average. What's the missing number?   i 3   ii 10   iii 13
   e The data in the table changes from year to year. Explain why.

3 Kenya is still quite poor. Look at the reasons given above. Pick out:
   a a historical reason
   b one reason to do with infrastructure (glossary?)
   c one reason to do with Kenya's climate

4 Explain how this holds Kenya back:
   a no electricity in many places
   b a shortage of teachers

5 China is building a new port at Lamu, to take huge container ships. How will this benefit:
   a Kenya?   b China?

7 Kenya

How much have you learned about Kenya? Let's see.

check ✓

A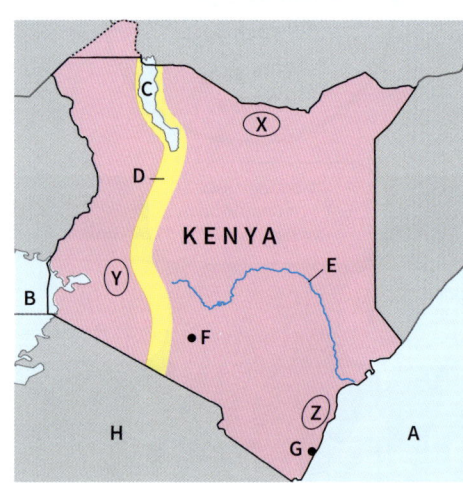

B International tourists arriving in Kenya

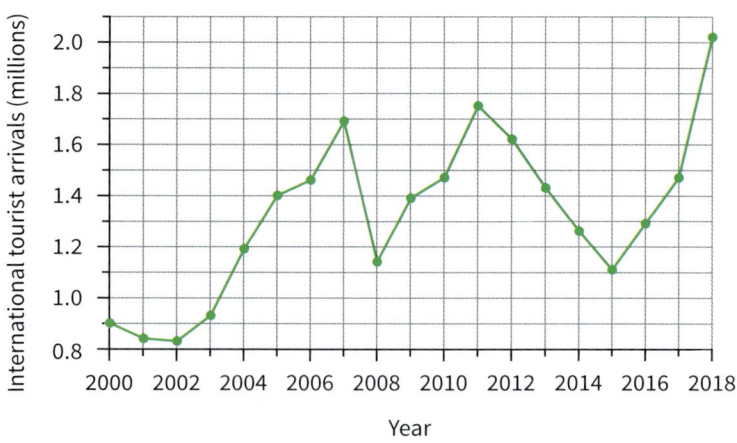

1 Map **A** shows Kenya.
  a Name the body of water labelled: **i** A **ii** B **iii** C
  b Identify the feature labelled **D** which runs through Kenya.
  c **E** is the longest river which lies fully in Kenya. Name it.
  d **i** Which letter refers to Nairobi, Kenya's capital?
     ii State three facts about this city.
  e Kenya's second largest city is also marked.
     i Name it.
     ii Give one fact about it.
  f Now look at the areas marked **X**, **Y** and **Z** on the map.
     i Which area is in Kenya's highlands?
     ii Which area has least rainfall?
     iii Which area is the least populated? Explain why.
     iv Give two reasons why area **Y** is good for growing crops.
     v What kind of farming are people likely to do in area **X**?
  g Identify the country labelled **H**, which borders Kenya.

2 Bar chart **C** shows the average monthly rainfall for the village of Korr in northern Kenya.
  a **i** Which month has most rain?
     ii About how much rain falls in that month?
  b In which months is the rainfall less than 10 millimetres?
  c How many rainy seasons does Korr have in a year?
     i none **ii** one **iii** two
     Give evidence to support your answer.
  d Korr is home to the Rendille people, who are pastoralists. Define the term *pastoralist*.
  e In which of these months might the Rendille have most difficulty in finding grazing for their animals?
     i December **ii** September **iii** April
     Explain your choice.
  f The rain is unreliable. Suppose that next year there's no rain in March – May. Predict one impact of this on the Rendille.
  g The total rainfall per year for Korr is likely to change in the future. Give a reason.

3 Graph **B** shows the number of tourists arriving in Kenya each year, in the period 2000 – 2018.
  a Give three reasons why tourists like Kenya.
  b Give one major reason why Kenya welcomes tourists.
  c Name one Kenyan national park which tourists visit on safari.
  d Describe one harmful impact of safari tourism.
  e *The number of tourists rose steadily from 2000 to 2018.*
     True or false? Give your evidence.
  f Which year on the graph was the peak year, for tourists?
  g How did these events in Kenya affect the number of tourists?
     i In 2007 an election led to violence between ethnic groups. Many people were killed.
     ii There were terrorist attacks in 2013, 2014, and 2015.
  h Suggest one reason why it is not wise for a country to rely too much on tourism.

4 *Although Kenya has a bright future, it faces many challenges. To what extent do you agree with this statement? Write at least five sentences.*

C Average rainfall in Korr, Kenya

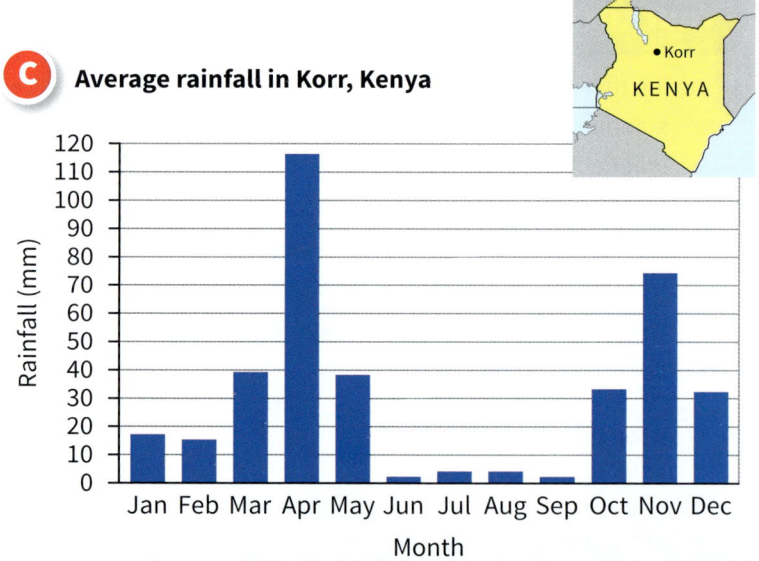

Command words: a summary

All through *geog.123* you'll meet questions with **command words** that tell you how to answer them. Later you'll meet them in exams too.

So it's good to get used to them.

Here's a summary of the command words to help you, in alphabetical order. The command words and their definitions are in red.

Assess
Weigh everything up and make a judgement.
For example assess the impact of an earthquake on a city. You must always include the evidence you based your judgement on.

Calculate
Do some maths, to get the answer!
Always put the unit in your answer. For example *5 km* or *11 people* or *15 days*. (Not just *5* or *11* or *15*.)

Compare
Identify what is the same, and different, about two things.
For example say which one is bigger. Always mention *both* things in your answer.

Copy and complete
Copy this, filling in all the blanks.
Fill in using the words and terms that make sense!

Define
Write down the meaning.
Keep your answer clear and simple.

Describe
Write a description.
For example describe what you see, or the steps in a process. You do not need to give reasons for anything.

Discuss
Look at something from different angles, and give key points about it.
For example you could give its good and bad points, or its benefits and drawbacks.

Draw
Draw!
For example draw a diagram, or sketch map, or bar chart, or line graph. Use a ruler for straight lines. Be accurate with bar charts and graphs. And try to be quick!

Evaluate
Judge how successful or worthwhile something is.
You should say what has been good and bad about it, *and* give your final opinion.

Examine
Look at each part and say how it contributes.
For example examine how different processes work together to form an oxbow lake.

Explain
Make something clear and easy to understand.
For example explain how a meander forms.

Give
Come up with an answer, from what you've learned.
Keep it clear and simple.

Identify
Pick out the thing, and give its name.
For example identify a landform on a map.

Justify
Give reasons to support the choice or decision you made.
For example give reasons why you agree with a statement.

Label
Add labels!
For example label a diagram. The aim is to make it clear and easy to understand. So keep labels short and simple!

Name
Write the name of the thing you are asked about.
Easy! You do not need to write a full sentence.

Outline
Set out the main points.
Stick to the main points. You don't need to give details.

State
Give the answer in clear terms.
State is often used in place of *Give* or *Identify* or even *Calculate* or *Count*. Be sure to answer clearly.

Suggest
Come up with a possible reason or plan.
Use your common sense!

To what extent .... ?
How much does it contribute, or how important / true is it?
Make a judgement, and give the evidence you based it on.

Ordnance Survey Symbols

ROADS AND PATHS 1: 25 000

- M1 or A6(M) — Motorway
- A35 — Dual carriageway
- A30 — Main road
- B3074 — Secondary road
- Narrow road with passing places
- Road under construction
- Road generally more than 4 m wide
- Road generally less than 4 m wide
- Other road, drive or track, fenced and unfenced
- Gradient: steeper than 1 in 5; 1 in 7 to 1 in 5
- Ferry; Ferry P – passenger only
- Path

PUBLIC RIGHTS OF WAY

1:25 000 / 1:50 000
- Footpath
- Bridleway
- Byway open to all traffic
- Restricted bridleway

RAILWAYS 1: 25 000

- Multiple track
- Single track
- Narrow gauge/Light rapid transit system
- Road over; road under; level crossing
- Cutting; tunnel; embankment
- Station, open to passengers; siding

BOUNDARIES 1: 50 000

- National
- District
- County, Unitary Authority, Metropolitan District or London Borough
- National Park

HEIGHTS/ROCK FEATURES 1: 50 000

- Contour lines
- · 144 Spot height to the nearest metre above sea level
- outcrop, cliff, scree

ABBREVIATIONS 1: 25 000 and 1: 50 000

PO / P	Post office	PC	Public convenience (rural areas)
PH	Public house	TH	Town Hall, Guildhall or equivalent
MS	Milestone	Sch	School
MP	Milepost	Coll	College
CH	Clubhouse	Mus	Museum
CG	Cattlegrid	Cemy	Cemetery
Fm	Farm	Hosp	Hospital

ANTIQUITIES 1: 25 000 and 1: 50 000

- VILLA Roman
- Castle Non-Roman
- Battlefield (with date)
- Visible earthwork

LAND FEATURES 1: 25 000 and/or 1: 50 000

- ruin — Buildings
- Public building
- Bus or coach station
- Place of Worship (current or former): with tower; with spire, minaret or dome; without such additions
- Chimney or tower
- Glass structure
- Heliport
- Triangulation pillar
- Mast
- Wind pump / wind turbine
- Windmill
- Graticule intersection
- Cutting, embankment
- Quarry
- Spoil heap, refuse tip or dump
- Coniferous wood
- Non-coniferous wood
- Mixed wood
- Orchard
- Park or ornamental ground
- Forestry Commission access land
- National Trust – always open
- National Trust, limited access, observe local signs
- National Trust for Scotland

WATER FEATURES 1: 25 000 and/or 1: 50 000

- Marsh or salting, Towpath, Lock, Slopes, Cliff, High water mark
- Aqueduct, Canal, Ford, Flat rock, Lighthouse (in use), Low water mark
- Lake, Weir, Normal tidal limit, Sand, Dunes, Lighthouse (disused), Beacon, Shingle
- Footbridge, Bridge, Mud
- Canal (dry)

TOURIST INFORMATION 1: 25 000 and/or 1: 50 000

- P Parking
- V Visitor centre
- i Information centre
- Recreation/leisure/sports centre
- Telephone
- Camp site / Caravan site
- Golf course or links
- Viewpoint
- PC Public convenience (toilet)
- Picnic site
- Pub/s
- Cathedral/Abbey
- Museum
- Castle/fort
- Building of historic interest
- English Heritage
- Garden
- Nature reserve
- Water activities
- Fishing
- Other tourist feature

© Crown copyright

Map of the world

— international boundary
• capital city

abbreviations
BELG.	BELGIUM
B-H.	BOSNIA-HERZEGOVINA
C.	CROATIA
CENT. AF. REP.	CENTRAL AFRICAN REPUBLIC
CZ.	CZECH REPUBLIC
F.	FYROM (Former Yugoslav Republic of Macedonia)
K.	KOSOVO
LITH.	LITHUANIA
MT.	MONTENEGRO
LUX.	LUXEMBOURG
NETH.	NETHERLANDS
S.	SLOVENIA
SE.	SERBIA
SL.	SLOVAKIA
SWITZ.	SWITZERLAND
U.A.E.	UNITED ARAB EMIRATES
U.S.A.	UNITED STATES OF AMERICA

Equatorial Scale 1: 95 000 000

Did you know?
- Earth is 4600 million years old.
- It weighs 6000 million million million tonnes.

The continents and oceans

Amazing – but true!
- Nearly 70% of Earth is covered by saltwater.
- Nearly 1/3 is covered by the Pacific Ocean.
- 10% of the land is covered by glaciers.
- 20% of the land is covered by deserts.

World champions
- Largest continent – Asia
- Longest river – The Nile, Africa
- Highest mountain on land – Everest, Nepal
- Highest mountain in the ocean – Mauna, Hawai
- Largest desert – Sahara, North Africa
- Largest ocean – Pacific

Did you know?
The world has:
- over 190 countries
- nearly 8 billion people
- over 6000 different languages

Glossary

A

abrasion – scraping away material

aerial photo – a photo taken from the air

altitude – the height of a place above sea level

aquifer – underground rock that holds a large amount of fresh water

arête – a sharp ridge, shaped by a glacier

asylum seeker – a person who flees to another country for safety, and asks to stay there

atmosphere – the layer of gas around Earth; at Earth's surface, we usually call it air

B

bedload – stones and other fragments that roll or bounce along a river bed

biome – a very large area with a similar climate throughout, and similar plants and animals

borehole – a hole bored into the ground, for example for pumping water from aquifers

C

capital city – where the government is based

cartographer – a person who creates maps

cartography – the study and practice of drawing maps

cash crops – crops that farmers grow to sell, rather than to use for themselves

climate – what the weather in a place is usually like, over the year

coloniser – a country that takes control of another country for its own benefit, and sends people to live there (settlers)

colony – a country controlled by another country, which has sent settlers there

command words – words used in questions, that tell you how to answer them

Commonwealth – an association of 53 member countries, including the UK; most of them were once British colonies

condense – to change from gas to liquid

confluence – where two rivers join

continent – one of Earth's great land masses; there are seven continents (see page 100)

contour line – line on a map joining places that are the same height above sea level

country – we humans have divided continents into political units called countries

corrie – a hollow where a glacier started; corries are also called cirques, and cwms

course (of a river) – the route a river takes on its journey from source to mouth

cross profile – a cross-section showing a river's channel and /or valley

D

densely populated – many people live there

deposit – to drop material; rivers deposit sediment as they approach the sea

desertification – where farmland or grazing land becomes like a desert, often through overuse

development – the process of change that goes on in a country, with the aim of improving people's lives

drought – there is less rain than usual, so there is not enough water for people's needs

drumlin – a long smooth hill shaped like the back of a spoon, created by a glacier

E

earthquake – the shaking of Earth's crust, caused by sudden rock movement

economic – about money and business

economy – all the business going on in a country; if more goods and services are being produced and sold, we say the economy is growing

economic migrants – people who move to a new place to find work, and to improve their standard of living

embankment – a bank of earth or concrete built on a river bank, to stop flooding

emigrant – a person who leaves his or her own country to settle in another country

emigrate – to leave your country in order to live in another country

Equator – an imaginary line around the middle of Earth (at 0° latitude)

equable – does not vary that much; the UK has an equable climate

erosion – the wearing away of rock, stones and soil by rivers, waves, wind or glaciers

erratic – a large rock that's different from those around it; it was carried there by a glacier

estuary – the wide mouth of a river, where the river water and sea water mix

evaporation – the change from liquid to gas

export – to sell goods and services to another country

F

flash flood – a sudden flood usually caused by a very heavy burst of rain

flood – an overflow of water from the river

floodplain – flat land around a river that gets flooded when the river overflows

fossil fuel – coal, oil, natural gas

freeze-thaw weathering – where water freezes in cracks in rock, making them bigger; eventually the rock breaks up

fresh water – the water found in rivers, lakes, wells, and streams; it is not salty

G

game reserve – an area of land set aside for wild animals, where they can be protected

GIS – geographic information system; it lets you display data on a map on a screen, to help you make decisions

glacier – a river of ice

glacial – to do with glaciers

glaciated – covered by glaciers, now or in the past

global warming – the rise in average temperatures around the world

gorge – a narrow valley with steep sides

GPS – global positioning system; radio waves from satellites are used to work out the latitude, longitude and altitude of a place

grazing – land with grass and other vegetation, where animals can feed

grid reference – a set of numbers, or numbers and letters, that tells you where to find something on a map

Gross National Income (GNI) per person – a measure of how well off people in a country are; you divide the country's total income for a year (in dollars) by the population

GNI per person (PPP) – GNI per person adjusted to let you compare countries fairly (because things cost less in some countries)

ground moraine – the material a glacier drops all over the ground when it melts

groundwater – rainwater that has soaked down through the ground and filled up the cracks in the rock below

H

hanging valley – a valley that hangs above a larger one; if it has a river, the water will pour down to the larger valley as a waterfall

I

ice age – a time when the average temperature of Earth was low, and glaciers spread

ice shelf – a sheet of ice that is attached to land, but floats on the ocean

immigrant – a person who moves here from another country, to live

impact – effect on someone or something

impermeable – does not let water pass through

independence – when a country stops being a colony, and governs itself

Industrial Revolution – the period (about 1760 – 1840) when many new machines were invented, and many factories built

inequality – when wealth and opportunity are not shared fairly among people

infiltration – soaking into the ground

Glossary

infrastructure – basic structures and services such as roads, water supply and sewage disposal that allow a place to run smoothly

invader – enters a country to attack it

interlocking spurs – ridges of land that jut out on opposite sides of a V-shaped valley, and interlock with each other like fingers

L

landform – a feature formed by erosion or deposition (for example a gorge)

lateral moraine – the material a glacier deposits along the sides of its route

latitude – how far a place is north or south of the Equator; it is measured in degrees

lava – melted rock from a volcano

longitude – how far a place is east or west of the Prime Meridian; it is measured in degrees

long profile – the side view of a river from source to mouth, showing how the slope changes

M

manufacturing – making things in factories

meander – a bend in a river

meltwater – water from a melting glacier

migrant – a person who moves to another part of the country, or another country, often just to work for a while

moraine – material deposited by a glacier

N

natural resource – a resource such as water, oil, soil, or sunshine, which occurs in nature and which we can make use of

nomad – a person who rears animals, and travels with them to find grazing

North Atlantic Drift – a warm current in the Atlantic Ocean; it keeps the weather on the west coast of Britain mild in winter

O

oxbow lake – a lake formed when a loop in a river gets cut off

OS maps – detailed maps of places produced by the Ordnance Survey

P

pastoralist – a farmer who breeds animals for a living, and may travel with them to find pasture

permeable – lets water soak through

physical feature – a natural feature on Earth's surface, such as a river, valley, or volcano

plan – a map of a small area (such as the school, or a room) drawn to scale

plateau – an area of fairly flat high land

plunge pool – a deep pool below a waterfall

population – the number of people living in a place

population density – the average number of people living in a place, per square kilometre

port – a place on the coast where ships load and unload cargo

precipitation – water falling from the sky (as rain, sleet, hail, snow)

prevailing winds – the ones that blow most often; in the UK they are south west winds (they blow *from* the south west)

Prime Meridian – an imaginary line that circles Earth from pole to pole; it is at 0° longitude

pyramidal peak – a sharp peak on a mountain, created by glacial erosion

R

raw material – a material that is used to make something else; examples are timber, crude oil, and the fibres from cotton plants

relief – how high or low the land is, compared with the surrounding land

refugee – a person who has been forced to flee from danger (for example from war)

ribbon lake – a long thin lake sitting in a trough that was created by glacial erosion

river basin – the land from which water drains into the river

rural area – an area that is mainly countryside, but may have villages and small towns

S

safari – a trip to see wild animals in their natural habitat, for example in Kenya

salt marsh – an area on the coast that is regularly flooded by sea water

satellite image – an image captured by one of the observation satellites that orbit Earth

savanna – has grassy plains with scattered trees

scale – the ratio of the distance on a map to the real distance

sediment – a layer of material (stones, sand and mud) deposited by a river

semi-desert – dry, with not much vegetation

services – activities people carry out to help other people; for example teaching them, or driving taxis for them

settlement – a place where people live; it could be a hamlet, village, town or city

sewage works – where the waste liquid from our homes is cleaned up, before it is put back in the river

sketch map – a simple map to show what a place is like, or how to get there; it is not drawn to scale

slum – an area of very poor housing

source – the starting point of a river

sparsely populated – not many live there

spot height – the exact height, in metres, at a spot on an OS map (look for a number)

standard of living – the level of wealth, goods and services available to people in an area

stereotyped – about the fixed opinions people may have, that do not reflect reality

striations – grooves in rock, caused by a glacier scraping the rock as it flowed over it

subsistence farmers – produce only enough to feed themselves and their families, with nothing extra to sell

suspension – small particles of rock and soil carried along in a river

sustainable – can be carried on without harm; for example sustainable tourism does not harm people or wildlife or the environment

T

tarn – lakes in corries are called tarns in the Lake District

terminal moraine – the ridge of material dropped at the front of a melting glacier

till – the mixture of rocks, stones, clay and sand that is carried by a glacier

tourist – visits a place where he or she does not live, for pleasure

trade – the buying and selling of goods and services between countries

transport – to carry things along

tributary – a river that flows into a larger one

tundra – a cold region where the ground is deeply frozen; only the surface thaws in summer, allowing small plants to grow

U

urban area – a built-up area (town or city)

U-shaped valley – a valley shaped like the letter U, carved out by a glacier

V

valley – low land, with higher land on each side; it was carved out by a river or glacier

volcano – a place where melted rock erupts

V-shaped valley – a valley shaped like the letter V, carved out by a river

W

water cycle – water evaporates from the sea, falls as rain, and returns to the sea in rivers

waterfall – where a river or stream flows over a steep drop

water vapour – water in gas form

watershed – the dividing line between one river basin and the next – usually a ridge of land

weather – the state of the atmosphere – for example how warm or wet it is

wildfire – a large fire spreading out of control in a wooded or grassy rural area

Index

A
aerial photo 8, 28
Africa 99 – 114
altitude 35, 36
Antarctica 39, 63, 75
Arctic 39
arête 67

B
biome 112
British Isles 42

C
channel (of river) 83
climate 47, 120
colonies in Africa 103
command words 14 – 19, 137
Commonwealth 57
compass 8, 23
confluence 82
continents 100
contour lines 36, 37
corrie 67
cross profile 83

D
deposition by glaciers 65
deposition by rivers 84
desert 112
developing 123
distance (on map) 32
drought 121
drumlins 71

E
embankments 96
Environment Agency 97
environmental geography 6
Equator 38
erosion by glaciers 64
erosion by rivers 84
erratics 71
estuary 82
European Union (EU) 57

F
flash flood 92
flood factors 93
floodplain 82
floods 92 – 97
flower farming in Kenya 130 – 131
fossil fuels 7, 75

G
GIS (geographic information system) 9
glacial landforms 65 – 71
glaciers 59 – 76
global warming 47, 121
globe 38
GNI per person (PPP) 134
Google Earth 9
gorge 4

GPS (global positioning system) 23
Great Britain 45
Great Rift Valley 118
grid references 30 – 31
groundwater 81

H
hanging valley 69
height (on OS map) 36
Himalayas 74
human geography 6

I
ice age 60
ice sheets 62
immigrant 48
inequality 7, 134
infiltration 81, 92
interlocking spurs 86
Ironbridge Gorge 12 – 15

K
Kenya 115 – 136
key (for map) 28

L
Lake District 72 – 73
landforms linked to glaciers 65, 66 – 71
landforms linked to rivers 86 – 87
latitude 38
Liverpool 26
London 54 – 55
longitude 38
long profile 82

M
manufacturing 129
maps 8, 22 – 23
meander 87
mental maps 26 – 27
migration to the UK 48
moraines 70
Mount Everest 74
Mount Kenya 119

N
Nairobi 126 – 127
nomads 128
North Atlantic Drift 46

O
OS map 34 – 35, 36, 73, 76, 95
OS map key 34, 138
oxbow lake 87

P
pastoralists 128
physical geography 6
plan 24
polluting (rivers) 89
population density 50, 108, 124
population pyramid 125

precipitation 80
Prime Meridian 38
Purley-on-Thames 95
pyramidal peak 67

R
rainfall in British Isles 47
rainforest 113
Republic of Ireland 44
ribbon lake 69
River Coquet 28, 35
River Mole valley 30, 33
rivers 77 – 98
River Thames 78 – 79, 94 – 95
rural area 51

S
safari 132
satellite image 9, 23
sat nav (satellite navigation) 23
savanna 113, 120, 121
scale 24, 25
sediment 84
semi-desert 112
services 52, 129
sketch map 28
spot heights 36
subsistence farmers 128
surface runoff 81

T
Temple Guiting 36
Thames Barrier 97
Thames Estuary 90 – 91
Thames Tideway 79
throughflow 81
tide 79
tourism 133
transport of material, by glaciers 64
transport of material, by rivers 84
tributary 82
tropics 39
tundra 60

U
United Kingdom 41 – 58
United Nations (UN) 57
urban area 51
U-shaped valley 68

V
volcano 119
V-shaped valley 86

W
Warkworth 28, 29, 35
water cycle 80 – 81
waterfall 86
watershed 82
water table 81
weather in the UK 46 – 47
woolly mammoth 60